play it
SAFE!

This book was originally published to accompany a BBC Continuing Education Department series, *Play It Safe!*, first broadcast in January 1992, presented by Anneka Rice and produced by Stephen Moss.

Other BBC Books to look out for:

Help Your Child With Maths
Help Your Child With Reading
Help Your Child With Science
Improve Your Child's IQ And Behaviour

THE COMPLETE GUIDE TO CHILD ACCIDENT PREVENTION

Dr Sara Levene of the
Child Accident Prevention Trust

BBC BOOKS

Published by BBC Books,
a division of BBC Enterprises Limited,
Woodlands, 80 Wood Lane, London W12 0TT

First published 1992

© Child Accident Prevention Trust 1992

The moral right of the author has been asserted

ISBN 0 563 36300 2

Illustrations © Kate Simunek 1991

Set in Linotron Souvenir Light and Medium
by Phoenix Photosetting, Chatham, Kent
Printed and bound in Great Britain by
Clays Ltd, St Ives plc
Cover printed by
Clays Ltd, St Ives plc

Contents

Contents

I would like to thank the following individuals and organisations for their useful comments on the text: St John Ambulance, British Red Cross Society, Mr A Chilton, Ms E Croucher, Mr C Downing, Dr R H Jackson, Dr H R M Hayes, Ms P Laidman, Prof. I B Pless, Miss B Sabey, Dr S Shepherd

Introduction

After John died we spent a lot of time talking, me, my wife and the other kids. Whose fault was it? We all knew about that window, but we never fixed it. It was all our faults.

This book is about keeping children safe – safe from accidents, that can hurt them badly or even kill them. In this chapter I am going to explain just how important that is. I am going to give some ideas about what mums, dads and anyone else looking after children can do, and how they can get other people to help. The rest of the book will give more details about being safe in particular places, and when particular things happen.

What is an accident?

An accident is something that happens when you don't expect it, which could lead to someone being hurt. You don't expect it; but that doesn't mean you were powerless to prevent it happening. Someone might even have been able to tell that it was likely to occur.

The bad news

Every accident to a child is bad news for a family and accidents to children – serious accidents – are common. Every day two or

On average, one child in five is taken to hospital after an accident every year.

three children die in accidents in the UK. That's more than 800 in a year. Excluding babies under a year old, more children die in accidents than in any other way. Twice as many die from accidents as from cancer, and three times as many die from accidents as die from leukaemia. More than 120 000 end up in hospital after an accident, and about one child in five – that's five or six in a class at school – will be seen at an Accident and Emergency department every year.

Children aren't grown-ups

Accidents to children aren't the same as accidents to grownups. That's because children aren't little adults. As they grow up, they don't just get bigger. They learn to do new things – exciting things like crawling, walking or riding a bicycle. They can take you by surprise. Suddenly they can do something for the first time. Like any beginner, they may be clumsy.

Children don't behave like grown-ups either. They want to play. They want to find out about the world. They do things for fun, like climbing trees or sliding down banisters.

Many accidents happen because kids don't concentrate. They rush about and fall over their own feet. They dash into the road. They gobble food and choke.

Children don't know how to keep themselves safe. They need other people to take care of them and that means you and me.

Accident-prone?

Are some kids more likely than others to have accidents? Are they accident-prone? It's difficult to prove that there is such a thing as an accident-prone person, but everybody has a story about a child who's always needing stitches. Some children are certainly more at risk.

Boys appear to have more accidents than girls. It seems that they get into more mischief. We don't know whether that's the way they are born, or whether we teach them to be like that, by saying things like 'boys will be boys'.

It seems that children from less well-off homes are often more prone to accidents than children from better-off families. They may not have a garden to play in. They may live in a rented home and the landlord won't put dangerous things right. There are more accidents in all families when they've just moved house, someone's not well, a single parent doesn't have enough help, or a parent is depressed or mentally ill. I hope I can give some safety ideas that are easy and cheap, as well as ideas about getting things changed.

Children have different accidents at different ages. Toddlers are at home and have accidents about the house. Older kids have accidents in the street. So they need looking after in different ways and at different ages.

The good news

The good news is that we don't have to take accidents lying down – we can put up a fight. We can do a lot to prevent kids being hurt.

We can stop accidents happening by, say, putting special tops on medicine bottles or driving slowly near homes. We can stop people being hurt even if there is an accident by wearing seat belts or bicycle helmets.

There are some things we can do straight away, things in our homes or gardens that are easy to change. People in power must also play their part. MPs need to make new laws. Planners need to put safe roads near homes. Architects need to put safe banisters on stairs. Designers need to make safe kettles. We can remind all of these people to keep kids safe. We can tell them what we want in our neighbourhoods and in the things we buy.

Most child safety is down to the people looking after children – parents, grandparents, child-minders, all sorts of carers. People like you! They can decide to be safe. They can get other people to 'think safety' too. I hope you will find many useful safety tips in this book.

Expecting accidents

A lot of people say accidents are a normal part of growing up. Of course everybody gets bumps and bruises. We've all got scars on our knees. But no one needs to have a bad accident, an accident bad enough to leave you in hospital – or worse. In a different way, accidents are a part of growing up, because the accidents a child is likely to have are part and parcel of how grown up they are.

Take falls. They are the most common accident for children of all ages. Let's take a child and follow it until it grows up.

WATCH OUT! Baby can crawl up the stairs and easily fall down.

How can a new-born baby fall? It can't do anything for itself, just wriggle its legs. The only way such a tiny baby could fall is if someone were to drop it.

After only a few months, the baby can wriggle and jiggle and roll itself over. Now it can fall if it is left on a changing table, or sitting on a bouncy chair on a work surface.

At nine or ten months, the baby can crawl. Now it can make accidents for itself. It can crawl up the stairs, but can't get down again easily. So it may fall down.

Once baby is walking, it will do a lot of falling and tumbling. Usually, it won't be seriously hurt. But what if it falls face first onto the corner of a table, or a coffee table with a glass top?

Growing up – expecting accidents

	at Birth	at 3 months	at 5 months
What baby can do	Lifts head for 3 seconds	Wriggles and kicks	Rolls over Reaches for things Holds things in hands Puts things into mouth
Falls	Dropping baby Falling with baby in back pack or carrier	Falling from high surface like changing table	
Cuts and bruises	Sharp toys left nearby		Playpen Sharp objects on the ground
Burns and scalds	Smoke detector in case of fire Bath – scald from hot water Hot feeds		Coiled kettle flex Reaching for hot cups and foods
Ingestion **Inhalation** **Suffocation**	Pillows, mattresses and duvets Plastic in crib Feeding with no adult watching	Dummies String and cord around neck	Playpen Choking on small objects
Poison			
Road safety	Baby seat for car travel		
Other	Bath – drowning		Babywalkers may lead to falls and burns

14

at 8 months	at 9 months	at 10 months	at 11 months	at 13 months
Crawls *Uses hands to open/shut fill/empty* *Sits with no help*	*Holds with fingers* *Pulls to stand*	*Walks holding furniture*	*Knows things still there when they are out of sight*	*Walks without help* *Crawls upstairs* *Can work knobs/ switches/dials*
Falling down stairs – use stairgate	Buggies, high chairs – *falling out – use* harness	*Falling over onto ground or onto things*	*Climbing and falling –* Stairgate	*Falls when climbing stairs* Stairgates
		Sharp edges on furniture		*Glass doors and windows* Safety film/glass
Fire-guard *Touching fires and radiators*	*Pulling hot things down*	*Hot cups* *Hot foods on tables*		*Falls onto fires* Fire-guards
			Opens lids, looks for things inside	
Child-resistant containers *Eating cleaners and medicines that are easy to get at*	Child-resistant catches Open cupboards *May eat or drink what it finds*		*Opens containers, eats what's in them* Child-resistant containers	*Opens household chemicals without child-resistant tops* Child-resistant containers
	Child car seat *for car travel*			

	at 18 months	*at 2 years*	*at 3 years*	*Pre-school*
What baby can do	Emptying/filling Copies grown-ups	Screws taps and tops Fascinated by smell/taste/texture	Stops putting things into mouth	Competent on stairs Climbs trees Manipulates cogs/nuts and bolts
Falls	Falls from windows Window locks	Falls climbing and playing Unscrews safety gates	Falls off garden toys Climbs out of cot	
Cuts and bruises	Sharp objects in drawers Child-resistant catches	Playground equipment e.g. swings	Sharp objects in home and garage	
Burns and scalds		Uses matches/lighters Turns taps on/off		
Ingestion *Inhalation* *Suffocation*	Plastic bags		Can still choke on peanuts	
Poison			May open child-resistant pots	
Road safety				Booster cushion for car travel
Other	Drowning in pond or pool			Drowning in open water

16

What about toddlers? They'd rather run than walk. What if they run and trip into a glass door? They are starting to climb too. Can they climb over the banisters or out of the window?

Even older kids who are steady on their feet can have nasty falls. They can tumble in the playground or from trees.

If you know your children – what they can do and what they like to do for fun – you can tell what sort of accident to expect.

Stopping accidents

Some parents keep an eagle eye on their kids. This can be a good way if you have plenty of people at home, like grand-parents or aunties. It doesn't work if there are several kids and just one grown-up, or if your family has to play somewhere far away and out of sight.

Some families like to have a lot of special safety equipment and to try to child-proof their homes. This is fine if you own your own home, so you can alter it as much as you like. You also need plenty of money and plenty of patience to fit everything, and to use it all the time. Some safety items are really useful, but we don't need to buy everything on the market.

Some people like to teach their kids to be safe by explaining everything to them. This can be a good or a bad thing. You need to know your child's mind. We just went through the way their body changes. Let's think about their mind too.

New-born babies can't understand anything you say. By a year they will know 'yes' and 'no', and be able to say a couple of words, but they are not going to understand if you tell them about safety. And they don't learn the lesson if they do have an accident. They might scald themselves by trying to drink dad-dy's tea one day, then do the same thing tomorrow.

17

WATCH OUT! If baby scalds himself with tea, he won't necessarily learn not to do it again.

Two year olds won't listen to a word you tell them. They'll just go off and do the opposite. At three, children are old enough to start to learn. They can get to grips with some ideas like 'hot' or 'sharp'. They can learn that the pavement is for people, and the road is for cars. A five year old can be told how to cross the road but can't be trusted to do so alone. At eight, children can remember what they have been told and be sensible long enough to cross a quiet road safely. A twelve year old can tell how fast cars are going and cross in traffic.

Getting it right

Every family will work out its own way to be safe. I hope this book helps you whichever you choose. There are three catch-phrases to remember:

1 Think safety

2 Know your child

3 Plan ahead

Think safety. Try to have safety in mind every time you think of the children. Buying for children – have you bought a safe pram, or one simply to match your colour scheme? Going out with the children – is it safe to put an extra child in the back seat? Keeping the children busy – is that toy with tiny pieces right for that two-year-old? You need to be on the ball in order to keep your kids safe.

Know your children. Know their bodies and the things they can do. Know their minds. How much can they learn and understand? What sort of children are they – good as gold, who listen and do as you say, or cheeky monkeys, who won't take any notice? This will help you work out what are the right safety measures for your home.

Plan ahead. Get started with safety before accidents are ready to happen. If it's something in the house, sort it out while baby is tiny: get that stair carpet fixed, or that window catch repaired. If it's something to buy, have one in plenty of time: get a baby car seat in time for baby's first ride home from hospital, get a stairgate before baby can crawl upstairs.

Quiz – can your child do it?

Here are some things children want to do, for fun or to help out in the house. Do you let your child do them? Think about how old a child needs to be to carry out each activity safely. Do the quiz with a friend. Do you both agree? There are no firm right or wrong answers. You know your own child best.

How old does a child have to be to do this safely?

1   Cross a quiet street alone

2   Make a cup of tea

3   Have a bath without a grown-up watching

4   Have a drink without a grown-up in the room

5   Play near water without grown-ups around

6   Ride a bicycle in a back street

7   Eat peanuts

8   Play with plastic bags

9   Walk downstairs without a grown-up

10   Climb trees

11   Play in the playground with other children and without grown-ups around

12   Use matches

13   Play with 'Lego'

Answers

1   *Eight years old. Children can't judge the speed of cars and cope with traffic until they're 11 or 12, but an eight-year-old can probably be trusted to wait on the pavement for traffic to clear. They may still dash into the road if they get excited. You can't trust younger children to stay still and wait. Children of two or three don't have any idea that cars are dangerous, and they can't play safely on the pavement. See page 67.*

2   *Ten years old. They're strong enough to pick up a kettle and a teapot, and they're not clumsy like a little child so they can pour the water without getting a scald. See page 26.*

3   *Four years old. They're old enough not to slip down and go under the water, and you can teach them not to fiddle with the taps. But that doesn't mean you can leave them alone completely. There needs to be someone about listening out. The bathroom door should be open and never locked. See page 47.*

4   *Eighteen months old. A baby this age can manage a teacher beaker by himself. Younger babies can choke on drinks or bottles. Even a toddler can choke on food and needs a grown-up there when he's eating. See page 32.*

5   *No age. Toddlers and babies can drown even in a paddling pool or a puddle. Bigger kids can drown outdoors. See pages 61, 110–12.*

6   *Eleven years old. Secondary school children are safe to use their bikes for school if they've had some cycle training. They need bright and reflective clothing, and a cycle helmet too. Most cycle accidents happen to youngsters between 10 and 14 so it's important to help them cycle safely. See pages 82–85.*

7   *Six years old. Peanuts can easily block the lung because they contain a special oil that makes the lung lining swell up. So*

they are more likely to make a child choke than a piece of plastic the same size. See pages 142–3.

8   No age. Plastic bags simply aren't toys. Children will be tempted to use them for dangerous dressing-up games. Six or seven year olds may be given plastic bags to keep things in, as long as they know not to use them for games, and there are no little brothers or sisters about.

9   Three years old. Two and three year olds can be killed falling on stairs. They need stairgates to keep them off the stairs and you walking with them, so that if they fall they bang into you. You can teach little kids to get down the stairs in special ways like bumping down on their bottoms, but they will forget to do this safely if there's no grown-up about. See pages 35–37.

10   Seven years old. Climbing trees is always dangerous but you can't always stop children having fun. Seven year olds can learn to test branches for strength. They can choose trees over grass not concrete. They can climb well unlike little children.

11   It depends on the playground. Toddlers won't be happy unless you're there. Some playgrounds are fine for older kids and some aren't safe even with a grown-up around because of broken glass or litter. See pages 89–98.

12   Seven years old. They're old enough to hold a match properly and to strike it without getting burnt. They can light their own birthday cake candles but they can't make a bonfire or light a fire in the hearth. Never leave matches where children can reach them. See page 39.

13   Any age – as long as they've got the right toy. 'Duplo' has big pieces which are safe and fun for babies and toddlers, but ordinary 'Lego' has tiny parts they can choke on. It carries a safety warning 'Not suitable for children under 36 months of age'. Always follow the safety warning. See page 102.

CHAPTER ONE

Room by Room

This chapter will tell you how to take a look at your home and find the danger spots for children. It will give you ideas about how to make things safer. It will not be simply a list of items to buy. It will also give you ideas about people who can help.

Let's take a tour around the rooms. Of course, many of the tips will work in several rooms. Just remember to 'think safety' wherever you are.

The kitchen

This is a good room to start with. Most mums will tell you they spend the whole day there. Babies and toddlers will spend a lot of time there near Mum. But remember there are plenty of hot, sharp and slippery things in the kitchen, so there's every chance for accidents to happen.

Bumps and falls

♦ You don't need to be a cleaning maniac. You don't want a grimy floor, but you don't want it wet and slippery. You don't want it polished like a TV advert either.

♦ See to splashes and spills straight away. You can slip in a puddle of juice as well as in a puddle of water.

Baby can sit safely in a high chair with a full safety harness.

♦ Keep clutter under control. All right, all kids have toys, and they like to spread them about. But make sure there is a cardboard carton or bin to scoop toys out of the way.

♦ Keep kids low down. No baby bouncers on tables or work surfaces.

♦ Sit kids safely. Always use a full safety harness in a high chair. A proper high chair you can fasten a child into is much safer than a pile of cushions or a plastic booster seat.

Burns

In my kitchen, the wall's all dark. My mum put something on to cook and me and my friends were out playing. My aunt was about to go to the toilet and she saw the smoke coming out of the kitchen.

Don't set the kitchen on fire. Keep tea cloths and towels away from hobs and gas rings – don't let them dangle down.

Beware of the chip pan. It's the most frequent cause of kitchen fires. Here's how to be safer:

1 Dry chips before you fry them.

2 Don't fill the pan with too much fat.

3 If blue smoke is coming off the oil, it's too hot. Turn off the heat and wait before you start to cook.

4 Turn off the heat when you go out of the room.

5 If the pan catches fire, turn off the stove or the hob.

6 Cover the pan with a lid or wet cloth (or a fire blanket). Never throw water over the flames.

7 Keep well away for half an hour. If you can't get the fire under the control get out of the kitchen, close the door and call the fire brigade.

If you do a lot of frying you might want to buy an electric deep-fat fryer.

Here are some general safety tips for the kitchen:

♦ Make sure you don't need to reach over the stove to get things. Above the stove is not a good place for cupboards or shelves.

♦ Make sure you put the kitchen matches well out of the way.

Like all menaces to kids, they need to be out of sight, high up (at least 5 feet above the floor) and even better, locked up, in a cupboard with a child safety-catch.

♦ Be aware of appliances that can get hot. Oven doors can get very hot. If you are lucky enough to buy a new oven, ask for one with doors that stay cool. Try to keep kids away from the oven door – a play-pen or a barrier can be a help. It is not just the oven that gets hot. The toaster or the washing machine on the hottest wash can also be extremely painful to touch.

♦ Be extra careful with the iron. Don't let the flex drag where someone can trip on it. Imagine you're a toddler sitting on the floor. What's at the top of that wire? It will be fun to pull down. Leave the iron somewhere safe to cool down when you're finished with it.

Scalds

Scott, age four, was dashing about the kitchen. It was spaghetti for tea – his favourite. He got so excited that he banged into his gran who was lifting the saucepan off the stove, and was scalded by the hot food.

Arjun, age one, was in the kitchen. His baby bath was on the floor. He climbed up on the bath to see what was on the worktop. He found a hot kettle and pulled it down, and was scalded on his arm and body.

It's surprising how far a little drop of hot water can go, and how much damage it can do. A child's skin is much more delicate than a grown-up's, and a baby's skin is even thinner and more sensitive. A small amount of water that would hurt your hand or part of your arm could scald the whole of a baby's head and face, or chest. So to stop scalds from happening:

Watch the kettle. Water in the kettle stays hot enough to scald a child for a whole half-hour after it is finished boiling. Don't put too much water in for a start. Many modern kettles have level indicators. This is an economy tip too! After you've made the drinks, pour out the water left in the kettle. Keep an eye on the kettle flex. Does it dangle from the worktop? Cut it short, or use a hook to keep it tucked away, or buy a special coiled flex. Dangling flexes are very tempting for toddlers to tug on.

Watch the hot drinks. Use a mug, not a cup. It's harder to tip over, and saves on the washing-up. Never, ever have a hot drink with a baby or toddler on your lap. They can reach for the cup or knock your hand and wind up with a tragic painful scald on their face and neck.

Mind that stove. Toddlers will want to reach up and grab saucepan handles if they are sticking over the front of the cooker. Use the back rings of the cooker and keep saucepan handles turned away. You might think about buying a cooker guard, but you don't need one. First, they don't do anything to stop toddlers poking their fingers through and touching the gas or electric rings. Next, they get quite hot themselves. Also, they're expensive. Lastly, they make cooking difficult. You can

---

Curly kettle leads can prevent children being scalded but they are an expensive extra, and it's just as good to keep the kettle flex really short. Most people aren't good enough with electrical things to do the job themselves and taking the lead to an electrician means yet another expense. A group of parents on one estate got together to do something about it. One Dad was an electrician and he had a session at the local clinic where people brought in their kettles and he shortened the flexes. He didn't mind helping out as he was part of the parents' group. Because they got together to think about safety, these families got something done.

---

hurt yourself trying to lift a big bowl of stew or soup over the top of a cooker guard.

Use place-mats. Table-cloths make the table look fancy, but it is easy for a child to grab them and pull a hot dinner or drink down on themselves. Know your child. If they are pulling themselves up to stand, or just curious, isn't that what they might do?

Be careful when you heat baby's feeds. A microwave isn't safe. The bottle may feel cool outside, but the milk inside may be hot enough to scald baby's mouth. Be careful with bottle warmers.

WATCH OUT! Toddlers can grab a table cloth and pull hot food down onto them.

They often have hot water inside that can spill. A jug of hot water can spill too, if that's what you use to get the bottle warm.

Sharp blades and cuts

♦ Store knives out of sight. They can do terrible damage. It's bad enough when children pick them up by accident, let alone when they use them in fighting games. Put them in a drawer with a child safety-catch if possible. Don't leave them out in blocks or on magnetic racks on the wall.

Are your children safe to help with getting food ready? Know your child: you can judge when they are safe with cutting and grating. Always:

♦ Keep kitchen gadgets out of the way. We all have too many things with sharp blades – blenders, processors, grinders. They don't all need to be out, primed for action and plugged in.

♦ Think safety when you buy. Can you poke your finger into sharp moving parts? Can you start the gadget without the lid on?

Poisons

Small children will eat anything they can lay their hands on. To help you plan ahead:

A one year old knows something is still around even when he or she can't see it, and will look inside a cupboard if they can get it open.

A two year old can unscrew a lid.

A five year old can often get the top off a pot with a special child-resistant lid.

All this helps us know what children are likely to swallow by accident. Babies under one will eat things that are lying about – soap, washing powder, whatever you forget to put away. Toddlers go for household chemicals like cleaners and bleach – they look inside kitchen cupboards. Older children swallow pills and medicines, things you have to look for in drawers and open packets or bottles to reach.

Teresa, age 11 months, was in her play-pen. She reached through the bars and ate a cigarette.

To help keep your child safe, **know what is dangerous,** like medicines from the doctor and most household cleaners. Alcohol can kill a toddler. Drain cleaner can burn their throat. Pills you buy from the chemist, like painkillers, and iron tablets some women have when they are pregnant – these things are dangerous too.

Store things carefully. The best place is out of sight, high up (more than 5 feet above the ground) and in a cupboard fitted with a child safety-catch. You may not have space to keep everything high up. If you have to keep things under the sink, fit a child safety-catch. They are cheap to buy. They are a bit tricky to fit, but once you've put them on they last a long time. They are safer than a lock and key because you have to use the catches every time. There is no way you can close the cupboard without them.

Keep things in the right bottles. If you swop bottles you can make terrible mistakes – think of weedkiller in a lemonade bottle. And if there is an accident, you need to take what was swallowed to hospital. Doctors use the bottle to find out what was in it and what harm it can do.

David, age 10, had been playing football and he was hot and thirsty. He saw a bottle of fizz and took a swig. But it was leftover weedkiller that Dad had poured into the wrong bottle.

Keep medicines in the kitchen. You don't leave kids alone in the kitchen for long so you'll spot them going for medicines. Medicines should still be shut away in a safe place. A locked medicine cupboard is a good idea, but a cupboard with a child safety-catch is more practical. The fridge isn't safe. It's easy to open, and what's inside is usually good to eat. When a medicine like a child's antibiotic says 'keep cool' it means don't leave it on the heater, not put it in the fridge. Fridge locks don't help much. Kids can open or unstick them for they are only held on by glue. Ask the pharmacist for advice about storing medicines when you collect a prescription.

Keep cleaners and medicines away from children with a child safety-catch.

Help children to know that medicines aren't sweeties. Don't use Smarties as magic pills when they hurt themselves. Don't pretend that the doctor's medicine is a sweetie.

Keep food and poison apart. Don't store household chemicals in the same cupboard as food stores. Children won't always know which is which.

Feeding baby

Saul was sitting on a pile of cushions on a chair. He was only 18 months old so he needed help to reach the table. He slipped off the chair onto the hard kitchen floor and broke his arm.

Here is some extra safety advice for meal times:

Watch your baby. Always be there at feed times. Never prop the bottle up to help baby feed alone. Be careful with solids too from the spoon stage to finger foods. Babies can easily choke.

Sit baby securely. Get a well-made high chair (look for BS5799 on the label). Make sure it is solid, with a wide base that is difficult to tip over. Make sure you have got all the nuts and bolts to stop it falling apart. Don't use chairs with split plastic if foam shows through – a piece of foam could be chewed or pulled off and cause choking.

Use those straps. Always use a proper safety harness in a high chair, not just the waist strap that comes with the chair. It is so easy for baby to tumble out.

Safety with gas

You don't want children to play with gas. They can start a fire or fill a room with gas so that it explodes.

Turn off the cooker. It's useful to keep gas turned off at the tap if it's somewhere handy you can reach. You can also turn off gas fires. Some gas taps have a key you can take out, to stop kids fiddling with them and turning them on by mistake. You can also try taking the knobs off the cooker if you've got a model where they pull off.

If you smell gas:

♦ don't use any electric switches – there might be a spark

♦ don't use any matches or cigarettes

♦ turn off the gas

♦ open the windows

♦ if you think it's a leak, not gas left on by accident, call British Gas. They are in the phone book under 'Gas'. They answer emergency calls 24 hours a day.

The hall and stairs

Let's move out of the kitchen towards the other rooms, looking around as we go.

The front door

Keep your kids in. Three year olds can use an ordinary door handle. If your door is easy to open, a toddler can get out. They can wander onto dangerous stairs or landings in blocks of flats, or into the street. So have a catch that is either too stiff or too tricky for the kids. If you can't get one fixed, use a bell or something jangly so you know they are at the door.

Keep fire out. Fires can spread easily if doorways onto common landings and halls are left open. Be extra careful to keep your flat door closed when you're at home.

Be able to get out easily. You may like to keep your door locked to keep burglars out. But you must have the key handy so that you can unlock it if you need to escape a fire.

Glass doors and dangerous glass

♦ Glass doors are a menace. When glass breaks, it makes long sharp pieces. Low glass (less than 2ft 6in. off the floor) is a menace to the under-fives who can run into it. Teenagers running and grabbing for the handle can put a fist through the door.

♦ For safer glass take out ordinary 'annealed' glass. Put in safety glass (toughened or laminated glass) instead. Remember patterned glass is never safety glass. In fact, it has weak spots and can be easy to break.

WATCH OUT! A toddler can open a door and wander off.

♦ Board up dangerous glass with hardboard.

♦ Cover dangerous glass with special plastic safety film. This is cheaper than buying new glass and doesn't show up when it's fitted. It lasts for years. It is a bit fiddly to put up, so read the instructions carefully.

♦ Stand something in front of low glass when it is in a window or patio door.

♦ Use stickers on the glass so it shows up and is clearly visible from far away.

34

◆ If you break any glass at home, wrap it up carefully when it goes into the bin. Have a lid that fits the bin properly to make it harder for kids to get at sharp tin lids or glass.

The hall

The hall is a good place to fit a smoke detector. See pages 38–40 for more tips on fire safety. Do you have rugs on the hall floor? They are easy to trip over. The worst sort of floor is polished wood with scattered rugs slipping about. Fitted carpets make the safest floor, if you can afford them.

The stairs

You might expect children to take the odd knock or bump on the stairs. But falls on the stairs can do a lot more damage than that. Every year several children die from tumbling down stairs. It is not just babies who are the victims, but two- and three-year-olds as well.

Keep clutter off the stairs. Don't leave a heap of things at the bottom to carry up. It's handy to have two baskets or cartons, one at the top and one at the bottom. Drop in the things you want to move up or down.

Keep small children off the stairs. You need a gate or barrier at the top in case someone upstairs trips up. You need one at the bottom too. Then a baby can't crawl up, or an older child wander up, and come down with a fall.

A barrier or a gate? A gate is easier to open. But you will need more than one. A barrier can be moved from place to place. You can even put it across a doorway in the bedroom or the kitchen. You must get extra sets of cups if you want to use a barrier at the top and bottom of the stairs.

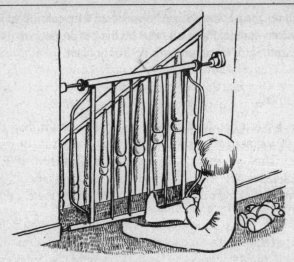

Barriers and gates help stop children falling on stairs.

Fix your barrier or gate the right way. As long as they have a label saying BS4125, they can all be safe. But you must use them properly. Use cups if they come with the gate; they hold the rubber feet safely onto the walls. Don't leave a gap more than 2 inches high under the barrier. Don't put them too far back on a stair or kids can use that stair to help climb over them.

Grown-ups be sensible. Open the barrier and don't climb over it. Otherwise you may have an accident with something that is supposed to keep you safe.

Are your banisters safe? Children can get through a 4-inch gap and fall down. They may use horizontal rails as a climbing frame to clamber over the top. You need to board up or cover railings at the top if they will enable kids to climb. Make sure your handrails are sturdy so they don't give way when someone's climbing the stairs.

Tyrone, age 14 months, was playing upstairs with some older children in a neighbour's house. He tried to get down to his mum but he fell through a gap in the banisters into the hall. He was killed. His dad had boarded up the banisters in his own home. They had the same dangerous design. His family didn't know they were left open in the house next door.

Check the stair carpet. Fitted carpet, if you can afford it, can help break a fall. But loose carpet, or worn slippery steps, can make matters worse.

Don't be in the dark. Have a night-light on landings, in case you need to get up, or for children who may wander about. Use a low watt bulb. Never cover a light with a cloth because the cloth can catch fire. If there are older people in the family, it's important to have a bright light at night, as their eyesight may be poor.

The sitting room

Let's move into one of the downstairs rooms. Where do you spend time as a family? You've probably got a sitting room, you may have a dining room and a TV room too.

Electricity

Homes usually have many electrical gadgets – TVs, stereos, video recorders and lamps. That means a lot of plugs and flexes are around.

♦ Keep sockets safe. Some babies like to play with them. Modern square-pin sockets are fairly safe. You can't poke your finger in the bottom holes unless you put something in the top holes too. But children can still burn fingers by pushing a knitting needle or a key into the top hole, then putting their fingers in the bottom. You can use plastic socket covers or move heavy furniture to block off sockets you don't use.

◆ Keep sockets clear. Put only one or two plugs into each socket. If you overload a socket, you may start a fire.

◆ Wire things safely. It's wonderful when electric gadgets come with plugs already fitted. Are you sure you know how to wire up a plug yourself? Remember to cut off the silver ends of the wires you often find when you buy something new as the silver solder doesn't make as good a contact as the clean copper wire. Look at the instructions – what fuse do you need? Most plugs come with 13 amp fuses so if the instructions are for a different type, take the 13 amp fuse out and put something else in.

◆ Check old flexes. Are they worn through? Is wire sticking out? Remember a child may put a wire into his mouth and be electrocuted.

◆ Tidy up the wires. Don't have wires trailing across the floor. If someone trips they can have a nasty bang. Worse, they can pull the wire loose and start a fire. Don't keep flexes under rugs and carpets. They might get worn through and cause a fire too.

◆ Unplug things at night, particularly the TV. They may catch fire otherwise.

◆ Think about keeping your home safer with an RCD (Residual Current Device). You can get an electrician to fit one at the fuse box. They cut off the electricity if something goes wrong. Or you can get a plug-in RCD for one gadget at a time. (They are useful for the lawn mower if you mow over the flex.)

Fire safety

Nick and Jason were four. They found some matches in the sitting room when they were poking about in Mum's handbag. They ended up setting fire to their clothes.

How do most fires start? Often they begin because smokers leave matches or cigarette ends lying about. These don't cause an instant blaze, but slowly build into a fire.

Be extra careful with matches and cigarettes. They can cause a terrible fire. That's another reason not to smoke! Empty all ashtrays before going to bed. Make sure you've cleared all old cigarette ends off foam settees and armchairs.

Foam furniture can be a menace. Furniture sold before 1988 contains old-fashioned foam which burns very hot and with a lot of smoke. It's the smoke that kills, not the fire. No one can afford to buy a new suite just for safety, but if you do need to change your furniture, look for the special safety label (see page 52). This symbol means that upholstered furniture has a fire-resistant filling, that most cover fabrics pass a test with a match and that fabric and filling material together pass a cigarette test. Be especially careful with second-hand furniture. It may not be that much of a bargain.

I was burning a bit of paper. The flames from the fire came up and burnt my finger.

WATCH OUT! Cigarettes and matches can start fires.

Have you got an open fireplace? Make sure it has a special safety fire-guard made to British Standard BS6539. It should be fixed to the wall so that a child can't pull it over. You will need a spark-guard too to stop sparks setting fire to the carpet when you're out of the room. Be careful with the fire-guard – don't use it as a clothes airer or as a handy table for drinks.

Keep the mantelpiece clear. You don't want kids climbing near the fire to take exciting things off the mantelpiece or to look at themselves in a mirror.

Bar the way to all fires, not just to open fireplaces. The bars on gas and electric fires don't stop children poking their fingers into somewhere hot. You need a proper fire-guard, labelled to a British Safety Standard. You can get guards that fix to the wall, and for heaters you move about like Calor Gas heaters. See page 150 for details of where to obtain a useful leaflet 'Keep them safe'. Try to put heaters out of the way where they won't get knocked over. Don't use a time switch to turn electric heaters on and off – an electrical fire might start when you're not around.

Shut the doors. Closing the door at night holds back a fire and helps stop it spreading round the house.

Choking and suffocation

There are many ways a child can choke, suffocate or strangle.

Little things and little children don't mix. Children under three like to put things in their mouths, in their ears, and up their noses. If they chew on something small it can go the wrong way to their lungs and choke them. As soon as the baby is playing on the floor, you must clear away all small objects. That means coins and buttons and pieces from toys. Make sure older children don't give baby anything small to chew.

Mind those pens. Older children don't choke very often, but they do love to chomp and chew on pen tops. Check their ball points and felt tips. Most will have special caps with air holes. That means if a schoolchild does get one in their windpipe, they will be able to breathe long enough to get to hospital and have it taken out.

Wind and pin away cords and flexes. A cord or a flex doesn't take long to strangle a little child. Children can strangle in the cords from blinds or curtains. So make sure all cords are wound up and pinned away. Don't leave cords or electric flexes dangling.

The bedrooms

Let's carry on upstairs to the bedrooms. They are places where children often play alone.

Mind those windows. Every year children climb out of windows and fall. They get very badly hurt and sometimes die. Children can squeeze through any gap bigger than 4 inches. Make sure your windows are difficult to reach. Don't put beds or chests in front of them so children can climb up. Fix some sort of safety-catch so they only open a little way. There are all sorts on the market, and some can be used on metal frames. If you need advice, a local locksmith will help.

Make sure the windows open. It is no good being locked inside if there is a fire and you need to escape. Keep the key to all locks handy – sellotaped to the frame is a good idea. You know where it is in an emergency, but it is not lying around for children to play with. That's why window bars aren't a good idea, unless they can be taken out quickly.

What's in your bedside drawer? That's a handy place to keep pills (especially birth control pills) and medicines, but it is not a safe one. Even some beauty items like perfume with alcohol or hairsprays can hurt children. Store all of these high up and out of sight, in a cupboard or drawer with a safety-catch. The same goes for matches, scissors and other sharp items.

Mind those bunk beds. They are great for children aged six and up to sleep in. They are not a good place to play, or for little ones to sleep in. Small children can fall out and hurt themselves badly. So keep them for older kids. New bunk beds (since 1987) have a rail on both sides, and the rails have safe gaps. If the gaps are the wrong size, a child's body may slip through but their heads stay stuck on the inside so they strangle. This has happened. If your bunk beds are old, or second-hand, check the gap between the top of the mattress and the bottom of the rails. It should be between 2½ and 3 inches.

Baby can sleep safely in this cot.

The nursery

This is where baby will spend a lot of time alone. Of course, babies may be sharing a room with you or with the other children. But you still need to be extra safe.

The cot

Rashu had just had his second birthday so he could stand and climb quite well. One night he piled up all his toys in one corner of the cot so he could climb out. His mum heard a terrible bang and found him unconscious on the floor.

A tiny baby can sleep in almost anything. You may want to use a carry-cot, a pram or a special crib. Make sure there is no pillow, or anything soft, like a lot of loose sheets underneath baby's face. Don't put baby down to sleep in a baby nest; it can get much too hot inside. For the big cot, look for BS1753 if you are buying new. For second-hand cots, make sure they come up to scratch. This means:

◆   the cot is deep enough to keep the baby from climbing out (at least 1ft 8in. between the top of the mattress and the top of the cot)

◆   there are no footholds in the sides or cut-outs in the ends which can help baby climb out, or trap his head, arms and legs

◆   the bar spaces are between 1 inch and 2½ inches, so baby can't trap his head

◆   the locks on the dropside work automatically, and the baby can't work them

◆   there are no transfers baby can chew on the inside

◆   the bars, the fastenings and the slats or springs on the base are in good condition.

Make sure the mattress is the right size for the cot. That means no gap more than 1½ inches anywhere round the mattress. A bigger gap could trap baby's head. The right mattress size may be marked on the cot or in the instructions. The mattress may also be marked with the right size for the cot. This is more important than whether the mattress has air holes or is covered in PVC – these two things don't really matter.

Don't use a pillow for babies under a year in case they suffocate. Don't use a duvet either; they can make baby much too hot, or baby can suffocate underneath. They are handy after the first year though.

Do you really need bumpers? They look pretty, but you need to take them out as soon as baby can sit up. Otherwise baby can use them to climb out. Keep any ties on bumpers very short – tie the bumpers on then cut off any spare.

If you're worried about baby wetting the mattress, use a properly fitted plastic or nylon sheet. Don't make do with the plastic that the mattress was wrapped up in or a plastic dustbin bag. Baby can suffocate in loose folds of plastic.

For toys in cots see page 102.

Once baby starts climbing out of the cot switch to a bed. Or leave the cot side down as far as it will go. You can fit a barrier across the bedroom door. And make sure you have got one across the stairs!

Changing baby

♦ The safest place to change baby is on the floor because there is no way he or she can fall. All you need is a plastic mat or a couple of old towels. You can also change babies on the cot. The bars help stop them rolling off.

♦ If you use a changing table or somewhere high up, make sure all the water, nappies and creams are near by. You can't leave baby for a second. Think ahead – baby will roll over soon, at least a week before you expect it.

♦ Baby can drown in a nappy bucket. So if you use terry nappies keep the bucket out of the way, perhaps in the bath. If you use disposables, remember baby can pull and chew the plastic tags and the wadding and choke. Make baby wear plastic pants on top of nappies.

♦ Keep piles of nappies out of the way. Stacks of nappies are a fire risk too, so mind matches and cigarettes nearby. By the way, the gel in super-absorbent nappies isn't poisonous.

♦ Beware of talc. If you use talc at nappy changes or at bathtime, it can make a cloud of dust that can choke a baby. It's safer not to use it for babies at all.

♦ Make sure you store all chemicals carefully. You may have sanitising powders for nappies, cleaners for changing mats, disinfectants for potties. The same rules apply – keep them out of sight, as high up as possible, and in a cupboard with a child safety-catch.

♦ A baby monitor can make you feel more relaxed. But it isn't necessary for safety's sake.

Warmth for baby

You want to keep your baby warm and snug. Remember:

♦ heaters must have guards so no one can touch the heater and get burnt (see page 40).

♦ keep the cot away from the heater in case baby reaches out to it or climbs out of the cot on top of it.

◆   overheating is also a danger to baby. You don't need extra heating in a centrally-heated house until the heating goes off when the adults go to bed. Don't wrap baby up too warm until your own bedtime, then tuck on an extra blanket.

The toilet and bathroom

At my Aunt Hettie's a wee girl got in the bath and it was too hot and she burnt herself.

Gino, age one, was playing near the bathroom. He helped himself to a drink. It was from a potty full of bleach that his mum had left to soak.

Be able to get out. You may want a lock for privacy, but you don't want small children to shut themselves inside. If you must have a lock, choose a stiff bolt and put it high up on the wall.

Store poisons safely. Again – out of sight, high up (more than 5 feet above the floor) and if possible in a cupboard with a safety-catch. What are bathroom poisons? They are bleach and other toilet cleaners, as well as your make-up or aftershave. They are also pills or medicines from the doctor or chemist. **Don't use block cleaners in the toilet**. Children often chew on them.

Store safely sharp things like razors, razor blades and nail scissors. Make sure they are out of reach. Don't have glass bottles around – they can easily get broken.

Don't get into hot water. Always run the cold tap first, then add hot water. Most of us run our hot water much too hot. This wastes money and is dangerous. Check your hot water tank, look for the thermostat and turn it down to 130°F (54°C). Then if a child does put his hand under the hot tap, or get into a hot bath, he won't be scalded immediately. Check the bath temperature with your elbow to see if it's comfortably warm.

Never leave young children in the bath alone or with an older child. A baby can drown fast. When the door bell rings, take no notice. When the phone rings, don't answer it, or wrap baby up in a towel and take him with you. Non-slip mats help to stop falls and bangs in the bath. Topple rings may give you a free hand to soap up baby, but they don't always stop baby slipping. They do not allow you to leave your baby alone. Don't rely on older children either.

Make bath time a peaceful time. If the evening is rushed for you, with hungry kids coming home from school, and someone arriving home for dinner, why bath baby then? Do it in the daytime, when you can be more relaxed, and it can get to be fun.

Fire alert

There is one danger that I want to go over in detail, and that is fire in the house. We have already thought about how to stop the fire starting – in the kitchen (see pages 25–29), in the sitting room (see pages 38–40), and in the bedroom (see page 41). Once a fire gets going in your home, you have only one minute – not five – not 10 – to get out. Most people die in fires because the smoke creeps up on them. They pass out peacefully while they're asleep. So fit at least one smoke detector. You can get them in nursery shops, children's shops and do-it-yourself shops. They are cheap and you need only one screw to put them up. They make a terrible din and will alert the whole family.

Fit at least one
smoke detector in
your home – you
only need one screw.

Put a smoke detector in the hall. If you have only one and your flat or house is all on one level, put it halfway between the living room and the bedrooms. If you have one smoke detector and the house has two storeys, put it in the hall above the bottom of the staircase. It's best to have one on every floor, if you can. With some brands you can wire the detectors together so they all go off at once, though they are more expensive and can be more of a nuisance. Look for the British Standard number BS5446 when you buy. **Check the batteries every week.** Like all safety gadgets, detectors are useful only if they are in working order. You can use an umbrella to press the check switch so you don't have to climb on a chair.

If there is a fire you need to get out fast. Don't wait for a fire to decide how to get out. There won't be time to think. Have an escape route planned. There should be two ways out of every room. One is the door. If you touch a door and it's hot, don't open it. If you go out of a room, shut the door behind you to hold back the flames. The other is a window that isn't locked or painted over. So make sure you can open a window if you have to. If you can't get it open, use something heavy to break the glass in an emergency. On **ground floors** and even **first floors,** you can go straight out. You do yourself less damage dropping out of the window than being trapped by flames. I do mean drop. Hang onto the ledge and slide out before you let go. You may be able to make a rope using sheets. You can put a blanket over the window frame to stop yourself getting cut if you have to break the window to get through. On **upper floors**, think about fitting a rope or ladder you can buy as a fire escape. Phone for help from someone else's home. Never go back inside.

Have a family escape plan. How will you get out of each particular room? How will you help people who can't walk? Where will you all meet up after the fire?

Tell children the two things they have to do immediately in the event of a fire. Firstly to shout, 'fire', and secondly to get out of the building as quickly as possible. Get all children over six to take care of themselves. Make sure they know, in advance, how to get out. Don't ask them to attend to anyone else. Only let adults help babies and toddlers out. Know where the little ones sleep, how to get to their rooms from your bedroom, and how to get out as quickly as possible.

Practise a fire drill. Do it in the daytime first. But remember, a real fire is black as the darkest night, so try it at night-time as well. Remember:

♦ Set off the smoke alarm.

♦ Shout, 'fire'.

♦ Order everyone to stop what they are doing and drop to the floor.

♦ Crawl to the designated exit for the room.

♦ Leave the pets and the teddy bears behind.

Except when the chip pan is on fire (see page 25) you usually won't have time to try to put out the flames. If you find a fire, get out of the room, shut the door, leave the house and phone for help from someone else's home. **Never go back inside.**

If you want extra advice about preventing fires or escaping, contact the fire prevention officer at your local fire brigade.

Getting things changed

I've given tips about what you can do yourself to make your home safe. But what if you live in a council house or rent your home from a private landlord? Here are some ideas:

Don't go it alone. You will find that people will listen to a group of you but not to just one person. Get together with other people in your block of flats or on your estate to form a residents' association. Get someone who's used to talking to officials, someone like your health visitor, to help. They might give you a hand with writing letters and fixing meetings. If you can get a residents' group together, find someone who's good at talking to put your points across.

If you live in council accommodation, try your housing department. They may change things for you. You can ask your health visitor and your doctor for letters saying that something is dangerous and has to be altered.

If the council won't do anything, or if you're renting privately, try the environmental health department. The town hall will give you their address. They can take landlords to court to get dangerous things changed. These can be private landlords, or the council. The environmental health officer may be busy and not able to help straight away. So make sure you get other local people and your doctor and health visitor on your side.

If you're still worried, and you can't get any action, try getting some publicity. What about inviting a local newspaper round to take some photos, or telling local radio about the problems?

In this chapter, I have tried not to mention too many safety gadgets, but there are some things you will want to buy. If you have trouble affording them, you may be able to get a loan from the social fund. Try your local social services department. Your health visitor may know about schemes which lend out second-hand equipment for as long as you need it.

Ten safety ideas that don't cost anything:

♦ Use the back rings of the cooker and keep pan handles turned to the back.

Mrs Khan lived in a council house and was worried about safety in her home. As she had three under-fives she obviously wanted it to be as safe as possible. She had an open fireplace and couldn't afford a fireguard. She made enquiries with her health visitor who knew about a scheme, run by the council, which lent out that sort of equipment to families. But the health visitor saw something else that worried her. Ali, age two, had a special game which involved swinging from the wires which poked out of the fuse box in the hall. He wouldn't stop no matter how often he was told to. Mrs Khan didn't know how to get the council to put it right and her English wasn't fluent enough for her to complain on her own. The health visitor told the environmental health officer from the town hall and they took on the council together. They both rang the maintenance department and explained the problem. A workman soon arrived to box in the wires. Mrs Khan had managed to make her home a safer place simply by looking for the right person to help her.

♦ Never leave a baby alone in the bath.

♦ Don't buy a babywalker.

♦ Turn down your hot water thermostat to 130°F (54°C).

♦ Keep the kettle flex tucked well back on the worktop.

♦ Clear the floor up so little kids don't choke on little bits and pieces and big kids and adults don't trip.

♦ Store medicines out of sight and high up in the kitchen.

♦ Change baby on the floor.

♦ Hold onto toddlers in the street.

♦ Don't smoke – or else put cigarettes out very carefully.

Safety symbols to look out for

Fire-resistant furniture (swing label and permanent label)
The furniture has passed the tests for fire-resistant filling, cigarette-resistant upholstery and match-resistant cover fabric.

Kitemark (British Standards Institution)
The item has been made to a British Standard and there are independent checks to make sure it comes up to scratch.

BEAB mark
This is found on electrical items; it means they come up to a standard on electrical safety, and they've passed independent checks.

Safety Mark (British Standards Institution)
This is on gas cookers and other gas items. It means that the item has been passed by British Gas and checked independently.

CE mark
On toys, the manufacturer claims they meet the European and British toy safety standards. On electrical items, it means that they are allowed by European law – it doesn't tell you much about safety.

Lion mark
This is found only on British-made toys. It means they come up to the same safety standards as CE mark toys; in fact, the toys will have the CE mark on them as well.

Toxic
This appears on chemical products that can disrupt the body's chemistry. The most harmful ones can't be sold on the DIY market.

Corrosive
This appears on chemicals that can burn into flesh. The most common one in the home is drain cleaner.

Harmful or irritant
These chemicals can have nasty effects but aren't as damaging as toxic chemicals.

Highly flammable
This label is on chemicals that catch fire easily. Some will have fumes that catch fire easily too.

CHAPTER TWO

Outside Your Home

The last chapter told you about the inside of your home, and how to make it safe. In this chapter, we will look round the outside of the house – the way in, the back yard, the garden. We'll look at blocks of flats too, the landings and the stairs. We'll be thinking more about play, and how to make that fun and safe, in chapter 4.

The front of the house

The front of the house will usually lead onto the street. You might have a drive where you can park the car.

My baby sister's only one and she's very safe but one time when we didn't have the wire over the gate she went into the road.

Keep children in. The front of the house is not a good place for the kids to play. The pavement in front of your home can be dangerous. It's easy for children to run into the road in front of a car. Parked cars hide kids from drivers passing by. Kids may hide underneath cars and be crushed when the drivers move off. Instead, make sure your children can't get out through your front door. (See page 33.) Don't let them rush out to say hello to daddy or visitors.

Fence it off. If you have a garden or yard at the front, have a fence with a gate all round if you can. Keep the fence low (you don't want to make a handy screen for burglars). You can put a gate across the drive, but this can be difficult.

Mind the bins. Keep dustbins out of the way, round the side of the house if you can. Dustbins are full of dirt and germs and hold sharp broken things. The big bins on wheels that many councils prefer are heavy and can hurt a child.

Danger in the driveway. Be extra careful when you park on the drive at home, especially if you reverse in. Several children get injured every year running out to say hello to Daddy only to be crushed by his car.

Lock the side door. Children can sneak out the front from the back garden. Burglars can get in this way too.

The garage

The garage is not a sensible place for the kids to be.

Don't run them over. Can your children get into the garage? If they play there on a rainy day, or wait there to greet you when you get home, you can squash them with the car. Be extra careful when you pull in. A bright light can help you spot the kids.

Keep them out. Lock outside garage doors, and the doors that go from a garage into the house. These keep children out – and burglars too.

Store it safely. Think about the sorts of things you keep in the garage. You might have:

♦ car tools (de-icer, a spare can of petrol)

♦ do-it-yourself tools (saws, knives, electrical gadgets)

♦ garden chemicals (weedkiller, antkiller, rat poisons).

Everything sharp or poisonous must be stored safely. That means out of sight, high up, and if you can manage it, in a cupboard that's locked or that has a child safety-catch (see pages 52–54 for labels on dangerous chemicals).

Don't pile up junk. There are many ways kids can hurt themselves if you have a stock of jumble and junk. They can climb to reach things and fall down. They can pull down shelves or heaps on top of themselves. There may be things that can catch fire if they find matches to play with. So if you store it, be sure you really need it.

The back yard or back garden

The back of your home may be a paved-over yard. This can be a good place for kids. A garden is also fun.

Can you see them? You can't spend all day watching the children, but you can check on them from time to time, say through the kitchen window.

Can they get out? It's no good thinking they're playing somewhere safe if there are broken bits of fence or holes in the wall. They may crawl through these. The football might land in the neighbour's yard and they'll be off after it. Board up all fences securely so you know where the kids are. Make sure any back gates are properly locked – a bolt high up (more than 5 feet above the ground) will be too difficult for the kids. So will a padlock. Most toddlers who drown do so in the garden pond next door. So fence in little kids, not just the older ones.

Can they fall? What is there for them to climb up? Try to stop them going high up in trees or on walls, especially if concrete is below. Children can fall over anything. They'll fall more often if

Keep an eye on children through the window, from time to time.

there are steps, small walls or split levels. They'll fall on wobbly paving slabs. Try to clear a flat area (a lawn or a yard) where they can play.

Is there rubble? Rubbish left from builders, and any other piles of junk or bricks, don't make good toys. They'll climb on the pile and fall. They'll bang themselves on the rubbish. They'll throw the junk around. Do try to clear out clutter if you want to leave the kids to play alone.

Is there firewood? You might be saving things for a garden bonfire. You might have old bits of wood from broken furniture or left over from building work. If you can make a bonfire, so can your children. Clear away anything they might use if they get hold of some matches.

Is there water? Children, especially toddlers, drown easily. They can go face down in a few inches of water. They can climb into a water butt or go head first into a bucket. Empty out all water collectors. Fill in puddles. If you give them water to play with, empty it as soon as the game is over. Watch the kids carefully if they're playing water games.

Mind the washing. You may have a washing line or a rotary airer in the yard. Children have got stuck in the lines and strangled. Make sure the lines are high up, well above the kids' heads.

Once you've made a trip to the tip, and cleared up your back yard or back garden, it can be an excellent place for kids to let off steam.

Extra problems in gardens

It often seems that a garden is a lot safer than a back yard. Sometimes that's true. It certainly hurts a lot less if you trip up on grass instead of concrete. But there can be extra risks in the garden too.

The garden shed

Like the garage, the garden shed is a risky place. You may have:

♦ a lawn mower

♦ garden tools like forks, shears or pruners

♦ weedkillers or other plant poisons

♦ slug pellets and other animal poisons

♦ steps children can use to climb up then fall down

♦ deck chairs that can squash fingers.

Your shed needs to be kept locked up. It's not a handy play-house in case it rains.

The greenhouse

The greenhouse can be risky from the inside if you use it as a store for tools, or for chemicals like weedkillers. So make sure the kids can't get inside. If the door is easy to open, put everything dangerous out of harm's way.

The greenhouse can be risky from the outside. That's because it's made of glass. Glass in the greenhouse is never safety glass, but ordinary (annealed) glass. If a child hits the roof with a football, that's a nuisance, but it won't do too much harm. If they kick the glass, or hit it with their bicycle, they can cut themselves. So cover the bottom of your greenhouse glass with plastic film, or board it over. Your plants will manage without the light on their roots. For more advice on glass, see pages 33–35.

WATCH OUT! Toddlers can easily drown in a pond.

Cold frames also have glass in them and they're low on the ground and easy for kids to break. Think about using plastic instead.

The garden pond

What's the most common way toddlers drown? In the garden pond. If you have children, think about it. Do you really need goldfish? Drain your pond. Fill it in in case it fills up when it rains. If you really can't bear to get rid of it, cover it over with chicken wire. Even better, put a 5-foot fence round it!

Dangerous plants

Gill, age four, was playing rabbits. She knew that rabbits eat grass and weeds so she stuffed herself with spurge, a garden weed. It tasted horrid but she didn't stop until her mouth and tongue started to blister. She was very sick but didn't come to any harm.

Most people have heard about certain dangerous plants, especially ones that are supposed to be poisonous. No children have died from eating **poisonous plants** since the turn of the century. I'm not saying there aren't poisonous plants in the garden, just that you have to eat an awful lot of them – like a soup plate of laburnum seeds – to come to any real harm. A child can get a tummy ache but that's all.

Plants can be dangerous for other reasons too. Kids can hurt themselves on **sharp plants,** like yukkas, roses and holly. Garden plants may carry tetanus, so you need to keep your child's jabs up to date, and go for immunisations when they are due.

Children find it hard to tell the difference between berries and fruits that are safe to eat – like bramble berries – and berries that can hurt them. So **teach them not to eat anything growing in the garden**, just in case.

Here are some plants you should ban from your garden:

♦ deadly nightshade. The purple shiny berries look delicious, but five or six are enough to cause serious poisoning. It's a weed, so pull it all up

♦ woody nightshade. This weed has shiny red berries, which are highly poisonous. Rip it up and burn it

♦ foxgloves. They have a heart poison in the flowers. Even the water is poisonous if you have them in a vase

♦ prickly bushes. Yukkas can be especially nasty. Some people cut off the ends of their leaves with scissors. Prickles and thorns might not hurt much on your skin, but they can do awful damage to your eye.

If your child does eat something out of the garden, do telephone your doctor or your local hospital for advice. They won't mind helping you. See pages 29–31 for more tips on poisoning.

Other garden poisons

Remember that slug pellets or other poisons to kill rats and mice can hurt kids too. See page 144 for first aid tips.

Helping in the garden

Your children will love to do whatever you do. If you are keen on the garden, they will want to help. Make sure you are around when they are gardening, so they use tools that are safe for them. Think about how old your children are. How sensible are they? You might let a three-year-old make a small hole, but only a teenager can mow the lawn.

Jobs your child can help with:

♦ planting seeds or bulbs (watch that small children don't eat them)

♦ raking leaves

♦ watering (always very popular)

♦ picking fruit.

Jobs they mustn't do:

♦ **Spraying chemicals.** Keep kids indoors during spraying. Many chemicals can get through the skin and poison.

♦ **Using weedkillers.** Especially the long lasting sorts that go on a patio or drive. You might like to kill weeds when the children are out of the house, so there is plenty of time for splashes to dry

♦ **Mowing the lawn.** This can be dangerous for kids even if they are only watching in case pebbles fly up and hit them in the face or eyes.

Living in a flat

If you live in a flat, you may have a balcony or communal landing.

Balconies

Make sure the way out to the balcony is safe. Balcony doors may have low glass. See pages 33–35 for tips on safety glass that won't break dangerously.

Can the children climb over the top? Children need a barrier at least 3ft 6in. all round the balcony. The railings shouldn't have

Make sure children can't climb over the balcony.

a gap bigger than 4 inches anywhere. Look at the pattern of the railings too. If they are horizontal or with a lot of curly cut-outs, there can be a ladder to help the children climb out. Block up dangerous gaps and railings with hardboard or chicken wire. You can make it easy for the children to climb over if you put plant tubs or chairs on the balcony. So keep the balcony clear.

Communal halls and landings

When a home is built, there are safety rules about banisters and stairs that help make the home safe for a child. These rules used not to apply to communal stairs and landings. That means the banisters in many flats won't hold a child back: they can get through the gaps or over the tops. The best way to be safe on stairs and landings is to keep the kids off them. Only let them play there when a grown-up or sensible older child is watching.

Communal gardens

The same safety tips for grass and gardens outside houses apply to flats. It is especially important to make sure kids can't get out if you can't see them while they play. Check the clotheslines. Children can strangle if the lines dangle low down so make sure they're above the kids' heads.

Getting things changed

If you think there are dangerous spots around the outside of your home, see page 49 for ideas about where to get help. If you are worried about poisons – chemicals or plants – ask your health visitor or doctor for help. They can get expert advice from a poisons centre. If you have dangerous chemicals at home you want to get rid of, you can get advice from the waste disposal department at the town hall.

CHAPTER THREE

The Street

In this chapter I'm going to explain about safety in the street. That's walking on the pavement and crossing the road, riding in the car, and for older children, riding bicycles, and playing in the streets.

Accidents in the street

The street is an extremely dangerous place for children. One or two children get run over and killed there every day. Children who survive are often badly injured. They may break their legs, which could mean weeks in hospital, or suffer head injuries, which can cause brain damage. Most kids who are killed or seriously hurt when they're walking or riding their bikes are knocked down by cars in side streets near their homes.

Why children get run over

Scientists have found that most children's car accidents occur when children run into the street without looking carefully. This happens even when they are with grown-ups, and holding hands. So some people might say that the accidents are entirely the fault of the child. But children can't take care of themselves. Drivers can help by taking extra care near kids. Also useful are low speed limits near homes so cars have to crawl past. If a car is going at only 15 mph, it won't do nearly as much harm if it hits a child as if it were travelling at 50 mph. Some countries have special streets round homes, with narrow roads and other ways to make it hard to drive fast. This cuts down accidents, and shows that drivers are partly to blame. But in Britain we don't usually have this kind of planning, so we need to take extra care to keep our kids safe.

Can your child cross the street?

I saw her on the pavement but I never thought a big girl like that would just run out into the road.

Know your children, and think about what they can do and understand. Many people feel that quite small children are safe to go out on their own to do shopping, or play with friends. Be careful you don't let your child take on too much.

Three-year-olds can learn that the pavement is safe, and the road is dangerous. They can't stop, look for cars, or cross any road on their own.

Five-year-olds can learn the Green Cross Code (see page 71) but it's not aimed at them. They don't understand most of it – they can't find a safe place to stop and can't cross safely.

An eight-year-old can understand the Green Cross Code and use it properly in a quiet street. They can't tell how fast cars are coming if there is traffic. They won't remember to use the code if they're playing with their mates. An eight year old can cross a quiet road on their own.

A 12-year-old can tell how fast a car is coming, and knows to wait if the car is travelling fast. Older children will cross roads safely as long as they're thinking about it. If they're playing, chatting or listening to a Walkman, they may forget. It's actually these older children who are killed or seriously hurt most often.

You are the best person to teach your child to cross safely. You have to be in the street to learn road safety properly, rather than in a classroom. Your child will learn by copying you from when they are tiny. Make sure they learn the right thing!

My brother had an accident. He was walking to see my dad and he was running and the car came. We had to take him to the hospital and my mum was crying.

Babies and toddlers

Babies will be pushed at first in a pram, buggy or push-chair. Be careful as you cross. **Don't push the chair into the road and then stop** to check for cars coming. Make sure baby is properly strapped in with a harness, rather than a push-chair waistband or bumper bar. You don't want him to wriggle out.

Toddlers walking on their own are hard to hold onto. They wriggle and jiggle and wander off. **It's much easier to use a harness and reins.** They may look like a puppy on a lead, but they have about the same amount of road sense. You may prefer a hand-to-hand strap that goes round your wrist. Buy one that's strong and short to keep your child close to you. **Keep toddlers off the road.** See page 33 for the front door. If you have a front garden, make sure it has a gate that your toddler can't open.

Amandeep, age three, was very excited. His big brother was getting married and there was a huge party at his house. He'd been allowed to stay up much later than usual. In fact, the grown-ups were so busy they'd forgotten all about him. He slipped out of the front door to look at the stars. He walked into the road as a taxi was pulling up – and was crushed.

Learning to cross alone

It's never too soon to teach your child the first steps in crossing the road. Just remember they won't be able to cross alone for a long time. So tell your three year old how cars can squash you. Teach them the difference between the pavement and the road. Help them learn the words for the Green Cross Code, like kerb, pavement and traffic. When you walk with your child, cross carefully. Tell them what you're doing – stopping at the kerb, looking and listening for traffic, waiting until it's safe to cross.

Show your child crossing places like zebra crossings, pelican crossings or bridges over dangerous roads. Use the crossings yourself, even if it means you have to walk a few extra yards. Set a good example so your children can copy you.

Ask your council's road safety officer if there is a traffic club or road safety programme for young children in your area. Traffic clubs are especially useful. They send children from three to five years old a pack every six months. If there isn't one in your area, ask your road safety officer and your council to set one up.

The Green Cross Code

The Green Cross Code has been specially designed to help children understand how to cross the road. It's not a rhyme or a jingle. Remember that children don't think the same way as grown-ups. Give them a jingle and they will use it like a magic spell. They say it three times and think the cars can't catch them even if they run into the road.

The code tells kids to:

♦ find a safe place to cross then stop

♦ stand on the pavement near the kerb

♦ look all round for traffic and listen

♦ if a car is coming, let it pass. Look all round again

♦ when there is no traffic near, walk straight across the road

♦ keep looking and listening for traffic while you cross.

The code is meant for seven- to nine-year-olds. Don't trust any child to stick to it. You've got to go over it with them and use it yourself every time you cross a road together.

A safe place to cross

Many childen find this bit of the code hard to understand. Help explain what these terms mean:

♦ a zebra crossing

♦ a pelican crossing

♦ an underpass

♦ a bridge over a road

♦ a big gap between parked cars, so you can see the road a long way in each direction.

A place that's not safe to cross

These days, with so many cars parked in the street it's difficult to find a gap between them when trying to cross the road. We grown-ups are used to going between the cars and looking out for traffic. But children are too small to do this properly and safely. They can't see over the cars. And drivers can't spot them. **Teach your child not to cross near parked cars.** If they have no choice because there are always a lot of cars parked nearby, teach your child to go to the outside edge of a car and stop. Teach them to look carefully for cars coming from both directions and only cross when there's no traffic near.

The Green Cross Code

1 Find a safe place to cross, then stop

2 Stand on the pavement near the kerb

3 Look all round for traffic and listen

4 If traffic is coming, let it pass

5 When there is no traffic near, walk straight across the road

6 Keep looking and listening for traffic while you cross

How to use a zebra crossing

1 Stop at the kerb at the crossing

2 When the traffic has stopped, walk across

3 Keep looking and listening while you cross

4 If there is an island in the middle, stop and wait again for traffic to stop

How to use a pelican crossing

1 Stop at the kerb at the crossing

2 Press the button and wait

3 When the steady 'green man' shows, check that the cars are stopping then walk across

4 Keep looking and listening while you cross

5 When the 'green man' flashes, stay on the pavement

6 If each side of the road has separate controls, press the button for the second part when you get to the middle and wait for the steady 'green man' again

Barriers and narrow roads can slow down traffic on your street.

A safe way to school

Jenny, age 12, was walking to school. She had to go along a busy main road but was a sensible girl. She saw her friend Trish on the other side. She dashed out without looking properly and a car ran into her. She had to be taken by helicopter to a special head injury emergency unit.

Your child may walk to school on his own, or with a bunch of friends. Do the walk with them three or four times. Show them safe crossing places. Make sure they know what underpasses are, and how all the crossings work. Try to work out a safe route. It might be a bit longer, but miss busy roads – and pass by the lollipop lady.

72

Be seen to be safe

Most children who get run over are hit in the afternoon and early evening, especially when it is dark and they're coming home from school. Children walking home from school are tired, hungry and in a big hurry, or playing with their mates. They're not concentrating on road safety.

Help drivers spot children. Give them something white or light coloured to wear. Put reflective strips on satchels or anoraks (you can often buy them with reflectors on already), or get them a reflective bicycle sash. You can also buy flashing armbands with cartoon figures, which kids will want to wear. Bright day-glo colours aren't good enough. They don't show up properly after dark.

The school bus

Drivers can help keep the school bus safe. They should remember that kids may run up to the bus or run off it quickly. School buses are safer if there's an adult or older child in charge. They can watch that kids don't lark about or try to get out. It's best if the driver operates all the doors. It also helps if the door is on the driver's nearside, at the front. That way the driver can make sure no one falls out, or gets trapped outside the bus by their bag or coat.

What the law has to say

The law can help to keep children safe. Laws and regulations use special words like 'footway' for 'pavement' and 'highway' for 'road'. Highway and road authorities have a duty under the law to make roads safe. They can:

- light the footway
- put up barriers or railings

♦ make traffic islands, subways and footbridges

♦ slow traffic

♦ ban or redirect traffic

Councils can arrange school crossing patrols. There are more ideas in the Department of Transport's leaflet *Children and Traffic*.

Getting things changed

The traffic round your home, or your children's school, may be fast and dangerous. You can help change this. First go to the traffic engineering department of your local council. They can carry out a survey and redesign the traffic flow around your whole area. Tell them what the problems are and how you think these could be put right. This way they'll be more likely to help.

If there's loads of traffic in your street, you don't know anybody who lives nearby. You like to get home, shut your front door and be safe.

If you've got cars cutting through your street when they should be on the main road, planners would call your street a rat run. No one wants to live in a rat run. The cheapest answer is to close off part of your street so people can't drive through. It just needs a simple barrier, which can let cyclists and fire engines through.

If the problem is big lorries, you can ask for a local lorry ban. The council may also need to make some parts of the road too narrow for lorries to get past.

If traffic is too fast ask the council to do something to slow it down. They could make the road narrow or put in speed

How one group did it

Anne lost her job and suddenly she was spending time at home. She realised her street was a nightmare with cars rushing past the homes and accidents happening every day. Drivers were using her street as a short cut. She made up her mind to do something about it. Five years later there were humps to slow the cars down and narrow places to stop lorries using the street. Here are some tips from Anne:

♦ **rope in all the neighbours.** It'll take time to get to know them; if there's lots of traffic in your street, you don't go outside the front door to meet people next door and you certainly don't cross the road to say hello

♦ **have someone in charge** who sticks with the campaign all the way through. They'll get to know the local officials, the local politics and the council's way of working

♦ **keep every letter.** Get one person to read the whole lot – they may remember some little item that will come in handy

♦ **get the local press on your side.** Short publicity stunts are useful for them and help get your message across

♦ **make a note of all accidents.** Take photos whenever you can – they are useful for publicity. Report every accident to the police and keep calling them out to make sure they make a written record.

Anne says it was terribly hard work to get their local scheme. But now the road is more like a big family with people knowing one another and helping one another in all kinds of ways.

humps. There are many other ways too. Because there are tight regulations on where speed humps can be placed, and some of the ideas can be expensive for councils, slowing cars down might be best in special areas like outside a school instead of for whole streets and neighbourhoods.

If you think there should be a crossing, you need expert help. The council decides whether crossings are needed using a special formula based on accidents, the number of cars and the number of people crossing the road. It's all in a leaflet (Department of Transport Advice Note/86 *Design Considerations for Pelican and Zebra Crossings*). See if the road safety engineering department at the town hall or the health visitor will help you get to grips with it. You should make a note of any accidents. The council is more likely to put up a crossing if you can give them a long list of accidents. Even if they won't put in a full crossing, they might build a traffic island, which can be helpful.

If you want to run a campaign, there are all sorts of things you need to do. Get hold of *Pedestrians – an action guide to your neighbourhood problems* (see page 150). Get advice from Friends of the Earth (see page 153). You will need to be tough but other parents have managed to get things changed to help keep kids safe in the street.

Safety in the car

Mrs Brown was driving the kids to school. The toddler was in his safety seat but she didn't have a booster cushion or a seat belt for her four year old, Sam. Someone ran in front of her and she swerved off the road into a lamppost. The toddler was fine though he got a terrible fright. Sam was thrown into the front of the car. He broke an arm and bashed his head.

How to run a campaign

It's not easy. You've got to get local people on your side as well as people in charge like the council.

Priorities – decide what's most important and start with that

Targeting – work out who's in charge and who has the power to do what you want. Focus your energies on this person

Facts – get facts together on paper with numbers, surveys and photographs

Report – produce something in plain English that looks professional, so it makes sense to councillors and the press

Publicity – don't be afraid to make a show and get the press on your side

The right seat for the right age

Babies – weighing up to 22 lbs (10 kg). Babies are best in a **special baby seat** (rearward facing infant carrier). These turn baby backwards in the car and let his whole back rest against the seat. They have straps you can adjust to hold baby in. If there's a bang, the force of the crash is spread over baby's back. Baby doesn't get a tremendous force on just one spot. These seats are fine even for the tiniest baby. You can hold his head with a rolled-up towel or a special cushion if it's wobbly. They won't hurt baby's back. Baby should use one from the first time he goes in the car on the way home from hospital. These seats need only an ordinary seat belt to hold them in. You can put baby in the back seat in one of these special carriers. You can also use it in the front. That can be handy if baby likes to watch you as you drive. It can also help you to keep an eye on baby. There are no yells from the back seat to worry about. It's also better for baby to sit near you if you've got toddlers who will pester baby in the back.

These baby seats aren't too expensive, and they can be useful for baby to sit in at home. But you don't need one for very long. Babies will outgrow it by the time they're about nine months old. There are many schemes to hire the seats. Ask at your antenatal clinic or ask your health visitor or road safety officer to find out if you can hire one locally.

Instead of a baby seat, you can use a **carry-cot**. It must have a cover fitted on properly and be held in by special carry-cot straps. The straps have to be bolted to the car, so you need to be handy or else go to a garage for help. A carry-cot takes up a lot of room and can be fitted only in the back. But it may be convenient if you use a carry-cot anyway for baby to sleep in. If you don't have a seat belt or carry-cot straps, the safest thing is to put baby's carry-cot on the floor behind the front seats. Place it with the head towards the middle of the car. Wedge it in firmly by pushing the front passenger seat back against it.

Don't hold baby on your lap. This is so dangerous that in the front seat it's against the law. Don't put baby inside a seat belt with you. You will squash him in a crash.

Baby can stay in an infant carrier as long as you can squeeze him in. You'll find his feet start to stick over the edge and you can't fit his legs in when you put the seat in the car. Or his head may begin to stick out over the front.

Toddlers weighing 20 to 40 lbs (9 to 18 kg). Toddlers are best placed in a **special child seat** with straps over their shoulders, round their waists and up between their legs (to stop them sliding down and out in a crash). You can get seats like this fixed into the car. It's best if they are fitted with four straps into the back of the car. You can also buy toddler seats that are fitted with an ordinary adult belt. This way you can move the chair from car to car.

What's the best brand to buy? From a safety point of view, there's not a lot in it. By law, all seats in the shops have to come up to safety standards. The best seat is the one you like, that fits your car well, that you will use on every single trip. It should be easy to keep clean and not too heavy to carry.

Is the seat fitted safely? If it fits with an adult belt, kneel on the seat to help tighten the seat belt properly. Make sure the child's straps are tight too. Choose a seat with straps that are easy to adjust. Don't let the seat belt buckle rub against the frame. If this happens with the adult three-point belt, put the seat in the middle of the back seat and use the lap belt. Even better, have the seat fitted in with a special fitting kit.

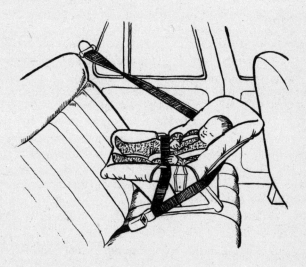

Use a seat belt from the very first trip.

Your child can stay in a special child seat until their eyes are level with the top of it. It's safest to stay in the seat as long as possible before moving on to a booster cushion.

Primary school children – from about four years old. You might think that older children are safe with an adult belt. But it's not designed for them. The shoulder part cuts across the neck. The buckle fits on top of their stomach. They don't have bony hips like a grown-up to hold the lap belt in place. In a crash, the buckle or lap strap can badly damage their liver or spleen.

Primary school children are better off using a **special booster cushion.** It's a hard cushion that's held in place by the seat belt. It holds the belt over the hips, so the lap strap doesn't do any damage. Booster cushions are cheap, east to fit, last for ages and you can move them from car to car. They are an excellent safety buy. An ordinary cushion isn't good enough – it slips out too easily.

You can also get booster seats that have a back support too, and two-way seats that start at the toddler stage. These are all safe for your child though they can be pricey.

Your child can carry on with a booster cushion until their head reaches the roof of the car. It's better to use an adult belt than nothing at all for kids this age. Never let children stand up in the back or poke their heads out of windows or sun roofs.

What the law has to say
By law, all new cars have seat belts in the front (since the 1960s) and in the back (since 1986). Older cars (since 1981) may have fitting points in the back, so you can put in seat belts.

Everyone in the front must use a seat belt or other strap that's right for their age. You can carry a child in the front only if they're strapped into a proper safety seat, like a baby carrier.

Everyone in the back must use a safety belt or child seat, if there's one there, and if it's right for them. That means you don't have to fit children's seats. You must put children four and up inside an adult seat belt if that's all you have. Even if the law doesn't make you buy a baby or child seat, your kids will be much safer if you do.

Too many people – not enough safety belts

Sometimes you need to squeeze a lot of people into the car – more than the car has seat belts for. Here are some ideas to help make your journey as safe as possible:

◆ Everyone in the front must belt up. Put the heaviest people, especially grown-ups, in the front. A grown-up can do a lot of damage to someone in the front, if they fly out of the back seat in a crash.

◆ Use all back seat belts. Secure the heaviest people first. Squeeze in other people. Don't use one belt for two. People squash one another in a crash.

◆ Put baby on the floor. A carry-cot with its cover on can be wedged down (see page 78) and so can an infant carrier. Wedge it down hard behind the seat. Use blankets or towels to help. Place baby so that his head is near the middle of the car and his feet near the door.

◆ Don't put children into the boot of any estate car or hatchback unless there are proper seats with belts fitted.

My child won't use his seat

Problem number one – a toddler fiddles with the seat buckle and lets himself out. Problem two – my child screams blue murder when he goes in his seat.

There aren't any easy answers to these problems. There are some things you can do to help:

◆ Get off to a good start. Use a seat for every journey from the first time your child goes in the car. Then they won't expect to travel without it.

◆ Have a rule 'no seat, no driving'. Stop the car if your child undoes the straps. Don't give in and let them get away with it.

◆ Fit a play tray which makes the buckle harder to reach.

◆ Keep the child busy with games like I-Spy (not reading in case it makes them sick).

◆ If it's the sort of seat that is held in by an adult seat belt, put the seat in the front so you can keep an eye on them, or as a treat for sitting nicely.

◆ What you can't do is have tough buckles that children can't undo, or hidden buckles a child can't find. Otherwise you might not be able to get the child out in a hurry if there's a crash.

On your bike

John was thrilled. He'd got a new BMX bike for his twelfth birthday. He couldn't wait to show his mates. He took it out as soon as he got home from school. He tried to turn right across the traffic. Perhaps he wasn't paying attention, perhaps he didn't know how to deal with traffic. He pulled across in front of a car and ended up with a badly cut head and a broken leg.

Children will want bicycles. They they will want to cycle in the road for fun or to get to school.

Start by buying a good bike. The kids will scream for something that's snazzy, fashionable and will do lots of tricks. But that kind of bike can have dangerous features like gear levers that come up between little boys' legs, or big saddles that encourage them to give rides to their friends. Ask the bike shop for advice when you buy new. Check the bike carefully if you buy second-hand.

Get some training for your child. The over-nines can receive proper bicycle training. The school may arrange it, or you can contact the road safety officer at your local town hall. Children will learn some basic rules of the road, and how to keep their bike in shape. Get the 'Highway code for young road users' for young cyclists to read (see page 150).

Use a bike only if it is in good condition. The brakes, especially, must work. A bike shop will help you if you buy the bike from them. A proper cycle training course will also help your child check the bike for themselves.

Get the right bike. You can't be a safe rider if the bike's too small or so big you can't reach the pedals properly.

Ride where it's safe. For young children that means the pavement or the park. It may be against the law but the police will usually turn a blind eye. Teach them to watch for cars turning into or out of driveways. They must always give way to pedestrians. Children from about nine up can go on local quiet roads (once they've had some training) but only secondary school children should ride in traffic.

Be careful on corners. Most accidents happen when cyclists are at junctions. It's important for children to know how to turn right and left. When they start, it's safer for children not to turn

Make sure young cyclists wear a bicycle safety helmet and bright or fluorescent clothing.

right; they should pull over to the left and walk across instead.

Be seen to be safe. Children on bikes need bright clothes by day – special fluorescent anoraks or waistcoats can be fun. For night-time (and remember that's four o'clock on the way home from school in winter), they need reflectors. These can be reflectors on the bike frame itself, on the pedals and spokes, and reflectors on their clothes, like stick-on strips and flashes. Reflecting waistcoats are also helpful.

Soften the blow if there is an accident. For minor bumps and falls, an anorak and trousers will help. In serious crashes, the worst thing that can happen is a bad blow to the head. For proper cycle safety, children need bicycle safety helmets. They have to be the right size, and fitted carefully. They may be fiddly at first but they're all you've got to cushion a bang. Racing or skateboard helmets aren't as good. They're not so strong, and can't protect the head as well.

Getting help and advice

You may not have money for all the safety equipment you want. Ask your antenatal clinic or health visitor if you can hire or borrow a baby car seat. Ask your road safety officer at the town hall if he can help teach kids about road safety. He will also know about cycle training schemes and where to buy helmets cheaply. Ask your school to do more safety teaching (see page 128). Get hold of the 'Highway Code for young road users' so your children can learn how to be safer (see page 150).

CHAPTER FOUR

Playing and Toys

They've got to play somewhere, haven't they?

This chapter is about children playing – keeping themselves busy and happy, and learning to do new things and get on with other people. I'll talk about good places to play, and safe equipment for playgrounds. I'll talk about good toys that are safe and fun at different ages. I'll tell you how to get things done – how to get help making a playground safe, how to complain if you buy a dangerous toy.

A safe place to play

When is a play area safe? Here are some ideas:

◆ It's fenced off to keep children in; there's a proper self-closing gate; the kids won't get out onto roads or wander off and get lost.

◆ Dogs are kept out. They can't bother the children or leave infected mess.

◆ Safe equipment has been put in, and it's checked regularly to make sure it's in sound condition (see page 89 for more on playground equipment).

♦ There's plenty of room between pieces of equipment, so children on swings don't hit children on paths, and children running from one piece to another don't collide.

♦ There's somewhere soft to fall like grass, bark chippings or a special rubberised surface.

♦ There's somewhere separate for the youngest children, or their equipment is at least in a special area.

♦ There is no water that children can reach other than a special water play area.

♦ There's a sign that tells you who owns the playground, in case you need to tell them about something dangerous or broken; where you can get help (like the nearest phone box); who is allowed to play there.

Safe play round your home

You may have a garden or back yard as part of your own home. If the children are going to play out the back:

Make sure they can't get out. Older children love to climb over fences or through holes. It might be for fun or to chase the football. Toddlers can wriggle through gaps.

Garden toys should be of good quality (look for BS5665 on the label) and set up properly. They should be checked every few weeks to make sure no nuts or bolts have come off. Remember that most garden toy accidents involve swings. Make sure swings aren't close to paths and other toys. Otherwise children get hit by the swings, or by other children jumping off them.

Emma, age one, was running in the garden while her big brother was playing on his swing. She ran near the swing and it banged into her head. She had a nasty bruise and cut her face, but she didn't need stitches.

Let children play in paddling pools only when you're there to watch. They can drown very quickly even when the water isn't deep. Empty the pool as soon as the kids have finished playing, and don't leave it out to fill up with rain.

Have something soft under climbing frames and slides they might fall off. Grass will do fine as long as it's watered. Dried up earth can be as hard as bricks.

Give toddlers extra help. They need special small slides and bucket seats and swings. They need you to help them on most toys and outdoor equipment.

Communal areas

My brother had to go to hospital. He fell on glass near the flats. He had 10 stitches.

Karen, age six, had cut her foot badly. There was only a tiny play area near her flats, and there was a pub over the road. People used to sling beer bottles over the fence and there was broken glass lying about. Karen's dad had a small flat and couldn't keep Karen in all day.

You may live in a block of flats or on an estate. There may be a special playground for children (see page 97) or somewhere the children always end up.

◆ Do the kids play alone? It needs to be fenced off, or far away from the road. Otherwise children will chase off into the street.

◆ Is it clear and clean? It's not safe for kids if there's junk like builders' rubbish or bits of broken glass. Or mess from dogs; this has all sorts of germs in it.

◆ Can the little children play somewhere separate? They will get hurt if they have to play close to big, boisterous kids.

WATCH OUT! Children can't play safely near junk and mess.

You can do something if you think your play area is bad. See page 96 for some ideas.

Playgrounds – safe equipment

There might be a proper playground near your house. Here are some things to look for:

♦ Play equipment must be safely made. Ask your local council if it comes up to British Standard BS5696. There can be other extra safety points too.

♦ You mustn't be able to fall more than 8 feet from climbing frames. They mustn't have cross bars to trap arms or legs.

♦ Slides should be set in slopes or banks. This way children can't fall off the edges. They must have handrails or side walls so kids can hold on to something as they slide down. It's best if they have ladders that aren't too steep with wide rungs, so children can easily climb up. There must be a long run-out and somewhere safe to land when a child comes off.

♦ Swings for young children should have special cradle seats. All swings should have rubber seats to cushion the impact if seats bang into people. Make sure the seats aren't too high, and that kids can sit on them properly. They shouldn't be too low, otherwise their legs will be squashed. It's best if swings are kept separate from other equipment.

♦ Roundabouts need to be low. They mustn't be able to go round so fast that a child running round the edge can't keep up and kids trying to get on can hurt themselves. They should be designed so that children can't get their hands or feet stuck underneath or between the rails.

♦ Playground equipment must be arranged sensibly. It's silly to put swings near the footpath so children swinging hit people walking past. It's dangerous to put pieces of equipment close together, so that children running from one piece to another knock into one another. It's a bad idea to put a slide in a sunny spot where the metal gets hot.

♦ Playground equipment must be fitted by experts, so pieces don't come out of the ground or collapse. Pieces should be checked regularly in case bits come off or get broken. You can check items yourself to make sure they are firmly held into the ground in concrete and are not rusted with missing bolts.

♦ It's much safer to have handrails round ladders, steps and platforms more than 2 feet above the ground.

♦ There must be some pieces of equipment for little kids, and some for older ones. That way they can all keep busy.

◆ There should be something soft and even to fall on. This could be bark chippings or special rubber sheets or tiles made out of soft materials.

◆ There should be a fence or cover to keep dogs out of sandpits, which should be disinfected and cleaned regularly – does your council do this?

Playgrounds – safe parents and children

When you take your children to a playground, you can help them to be safe. Make sure they're in sensible clothes. Things that dangle like scarves and anorak toggles can get trapped in slides and other pieces of equipment. So can hoods on anoraks or sloppy tops.

Little children:

◆ watch what they're doing

◆ teach them how to use the equipment

◆ hold them on when they're wobbly

◆ teach them to sit properly, and not jump about

◆ teach them to keep out of the way of bigger children playing lively games

◆ keep an eye on them if there's water, say a pond or paddling pool.

Older children:

◆ check they know how to use the equipment

◆ keep them away from little kids

◆ don't encourage wild games and dares

A good playground is both fun and safe.

♦ don't encourage using equipment in the wrong way, for instance, using the slide to see who can jump from the highest place.

Tell them to:

♦ look out for moving things like swings

♦ hold on tight when they're climbing

♦ tell grown-ups about anything dangerous or broken.

Surinda, age seven, fell off a chain swing. It was on concrete in the play area of a block of flats. Her head was cut and bruised.

Elroy, age 10, was playing with his mates in the playground. They were seeing who could jump furthest off the top of the slide. Elroy jumped a long way but fell as he landed and hit his hand. His wrist was broken and he had to be put into plaster.

There are ways to use each bit of equipment safely, but children do need to experiment and try things out. You don't want loads of rules in a playground that frighten the kids off. So here are some ideas for using equipment really carefully – but make sure your children have fun too.

On slides:

♦ slide sitting up, not backwards, lying down or head first

♦ no jumping off the platform

♦ no pushing or shoving on the ladder or on the platform

♦ no turning round on the ladder or on the platform

♦ no running up and down the slide.

On swings:

♦ keep away from moving swings

♦ no jumping off the swing

♦ no climbing up the swings or the frame

♦ no running underneath the swing.

On climbing frames:

♦ take turns instead of pushing or shoving

♦ no jumping from high up, even if there's something soft underneath to cushion you.

In sandpits:

◆ no throwing sand.

On roundabouts:

◆ take turns, particularly big kids

◆ hold on tight to the handles

◆ no crawling underneath

◆ no dangling arms and legs over the edge.

On see-saws:

◆ no getting off when one child's high up

◆ no crawling up the plank

◆ no hard bouncing to throw the other child off

◆ no touching the springs on rockers.

Dogs:

◆ playgrounds and dogs don't mix. Don't take your pets for walks near children's playgrounds.

Adventure playgrounds

Adventure playgrounds are special playgrounds where children can have excitement and fun. They can test themselves out, by climbing up high, by using ropes, ladders and nets. Most adventure playgrounds give grown-ups a fright. But they are great places for kids. Even big, tough boys like them. **Adventure playgrounds keep children out of trouble.** An adult play-worker is there at all times. They can help kids to build, explore and learn to get on with others. They can stop kids being too brave. They can make sure bigger kids don't go wild, or bully little ones. I'm not saying no one ever gets hurt. Some kids do get bangs and bruises and the odd broken arm. But the playgrounds keep kids away from really dangerous places where they could get very badly hurt.

A dangerous place to play

Some places, like building sites, just aren't meant to be used as playgrounds. They are much too dangerous. Make sure your kids understand the problems. If you think children can get into these places too easily, see page 97 for ideas on how to get things changed. Dangerous areas to play are:

Adventure playgrounds keep kids out of trouble and they're more fun to play in.

◆ In the road. Kids get hurt by cars driving past or parking. They may even use cars for games like running across the tops of them.

Jim's ball went under a car so he crawled under to get it. The driver got in and didn't notice him. Luckily Jim shouted out before he drove off.

◆ On building sites. Kids can climb and fall down piles of rubble, and into trenches where they may get trapped or drown. They may find dangerous chemicals.

Billy and his friends had learnt about safety at school. That's why the three of them were excited when they found the cans marked 'flammable' on the building site. They brought them home to have a bonfire in the garden shed. Billy was burnt to death. His mates were scarred for life.

◆ Quarries or gravel pits. There can be loose piles of sand or gravel, which can crush a child. Children may drown in pools or puddles of water.

◆ Close to water. It's one thing to let teenagers go fishing in a group. It's another for young children to walk or play near water, or for older children to use a canal bank as a football pitch. Teach your children to keep away from water.

Getting help and advice

Many local councils will have someone responsible for play. They should be able to give advice on what's available in your area, and what playgrounds are suitable for different ages. There are national associations like the National Playing Fields Association and the National Children's Play and Recreation Unit, and local and county playing fields associations (see page 154).

Getting things changed

We've heard about ideas for good places to play, but also about many dangers. What can you do to get problems solved?

If you see a playground that's dangerous, complain to the owners. It doesn't matter who they are: the council, the pub, the garden centre, the shopping centre. Make sure you report it if your child is hurt. Not enough people complain so kids go on getting hurt. Don't be afraid to get advice – the National Playing Fields Association may be able to help.

If you need a local play area, see if your council can help. Don't go it alone. Get your local tenants' association group to help. Get the local press on your side. Someone used to dealing with officials like your health visitor, a youth worker or a social worker, might help you to organise your campaign. Children of all ages need the right sort of place to play – doorstep play areas for toddlers, local play for primary school kids, neighbourhood play areas for older children. If people like you don't ask, planners and the council will forget about the local children.

Safe Play Campaign

A health visitor organised a parents' group on an estate. To begin with, they talked about feeding and sleeping and other worries with babies, then they started to talk about safety. The parents wanted to have a safe place for the little children to play. They found a local space that was supposed to be a recreation area. There was hard concrete and broken glass, and it needed to be cleared and tidied. They wrote to their local councillor and got up a petition. They went to listen in at a council meeting. They met some council officers who came round to have a look. The parents weren't afraid to have their say. It took ages but in the end, the council found the money to clear the ground and build a simple playground for little kids.

If you think the traffic is too dangerous round your home, complain to the council traffic department. They may be able to do something like putting bumps on the road to slow cars down. Get everyone in your street together. Ask the local newspaper to help (see page 75).

If your kids can get onto a building site, complain to the owners. They may be breaking the law. If your kids can get onto other dangerous spots, complain. Make a big fuss.

Toys and safety

We all expect toys to be extra safe. After all, they're for children. Kids often play with toys on their own, to keep themselves busy when nobody else is about. But every Christmas we see dangerous toys on television: teddy bears that catch fire, or dolls with nasty spikes inside their necks. The good news is that most toys are safe. Some safety rules for toys have the law behind them. These rules are for the whole European Community. They mean that nearly every toy you buy should be safe. But children always take us by surprise. They may do something unexpected with their toys. So people who make and test toys do miss dangers sometimes. You can check toys yourself to help your family be safe.

Rules for toys

The safety standard for toys (BS5665 or EN71) covers all sorts of safety points. If a toy comes up to scratch, it will have the CE mark on the label. Toys made in Britain will also have the lion mark (see page 53). Some of the areas covered in the rules are:

♦ dangerous paints or dyes. They don't allow chemicals which can do harm, like lead

♦ sharp points and edges. They're all banned

♦ small bits that could choke a child. They're not allowed for young children

♦ toys that fall over

♦ fur fabric that can catch fire

♦ other materials that can be a fire risk

♦ electric toys that plug in. They must always use transformers or batteries.

Why children have accidents with toys

Usually, it's nothing to do with the toy itself. Because of the toy safety rules, and because most manufacturers do care about kids, most toys are safe. But there are two big problems. First,

WATCH OUT! Don't let toddlers get hold of older children's toys.

people fall over the toys if kids leave them lying about. So have a toy box, basket, or bin where kids can put their toys. Second, children get hold of the wrong toys. A baby or toddler is almost certain to put a tiny building brick up their nose, in their ear or in their mouth. That's not being naughty. That's what you'd expect them to do. But they can choke this way. Toddlers can have accidents on toys like trundle trucks; if they sit on a sit-upon toy, but they're still wobbly, they're going to wobble off.

Ben, age three, fell off his rocking horse. He banged his head on the wall and cut it. There was a lot of blood but luckily his mum knew what to do. She pressed hard until the bleeding stopped and left the scab alone until it dropped off. Ben still has a bald spot under his hair.

John, age eight months, was sitting on his fire engine. He fell off and cut his nose.

Tips for toy safety

◆ Buy from a good toy shop. It could be a special toy shop, a well-known high street chain, or even a market stall, as long as it's there every week, and is not a fly-by-night.

◆ Check the toy yourself. Ask to see one out of the box. Make sure it is sturdy and solid, and doesn't break apart.

◆ Read the label. Check for the CE mark. This should be a guarantee of safety. Check for other safety messages: 'Not suitable for children under 36 months of age'. This means the toy will have small pieces that can choke a little child. It isn't safe for younger children even if they're very bright. 'Novelty. Not suitable for children.' Do not buy this as a toy. It could easily contain poison or sharp edges. It isn't made for children to play with. 'Play age five to seven'; 'Suggested age three to six'. These are advice messages, to help you to choose, not warnings.

Tom, age two, was fiddling with his brother's toys. He found a load of building bricks and stuck one up his nose. His Dad couldn't reach it and he had to go to hospital for the doctor to pull it out.

◆ Know your child. Always read the safety messages. You don't always have to follow the advice messages on the packet. Remember a babyish toy won't be much fun, and a difficult toy will just put them off.

◆ Get a toy box. Make sure it doesn't have a lid that slams down with a spring, and there are air holes in case someone gets trapped inside. A plastic stacking box, or even a cardboard carton will do fine.

◆ Check your toys. Broken toys can cause accidents, so check the toy box and clear out broken ones. Don't pass them on to the jumble – that's just handing on your accident to someone else.

◆ Don't mix batteries. Change all the batteries at once, and put all the same kind in together. Otherwise the strong batteries make the weak ones very hot.

The right toy

Furry toys. Toys with shaggy fur or a lot of hair aren't good for small children. If they suck the toy, they can get a mouthful of fur, which might choke them. Little children won't really play with furry toys either. Toys with fur are safe for children over a year old, and they're fun for children aged about three and up.

Toy guns. I won't talk about whether guns are good for boys. Do guns make them rougher? Do guns make them wild? All I know is that little boys love guns. If you don't give them a gun, they'll make one with a banana or building bricks anyway. Guns can be dangerous if they shoot things out. Check any gun that shoots. Are the bullets or darts that come with it really soft? Make sure they don't have pointed ends, even if they're plastic. Make sure they can't use the gun to shoot any other bits they've poked down the barrel. Teach children never to point a gun in another person's face. A pellet close up can badly damage someone's eye.

Guns can also be dangerous if they're especially noisy. It's hard to tell how much noise could make a child deaf later on. So teach children not to shoot noisy guns next to people's ears. When you shop for toys, try to buy quiet ones without too many electronic sound effects.

Building blocks like Lego. These are carefully made and well thought out. The special large pieces are ideal for toddlers. The little pieces are great for older children. They're good fun, they teach children how to use their imagination and their hands, and they last for years. But remember that they are dangerous for the under-threes, in case they chew on them and choke.

Toys for cots. If you tie anything onto the cot, it must have a short string – never longer than 1 foot. Otherwise it could strangle a child. Tie it on firmly. If you tie a toy across the top of the cot, you must take it out when baby is old enough to get on his hands and knees and reach it. As soon as baby can pull himself up to stand, take off toys and activity centres fixed to the side of the cot. They could give him a foothold to climb out. Don't stick pictures up on the inside of the cot. Baby might chew on them and then choke.

Nigel, age 12, was skateboarding along the pavement when he hit a bump in the paving stones. He came off with a crash. Luckily he was wearing some pads on his knees and elbows. He had a helmet on but it wasn't a proper safety helmet. It fell off and he cut and bruised his head.

Skateboards. Skateboards are a lot of fun with the right gear. Elbow and knee pads help prevent knocks and bangs. But most helmets are a gimmick, and do little to stop nasty bangs on the head. You might do best to give your child a proper cycle helmet (see page 84) instead.

Darts. Darts aren't toys. Metal darts can get into a child's eye. Don't let children play with ordinary darts.

Chemistry sets. They are meant for children eight and up, and they're probably best for over-12s. They're for children to use alone, but somewhere near you so you can help out if you're needed. Try to find space in the kitchen, and don't let the children mess about with them in their rooms. Make sure they read all the instructions carefully. If they need safety goggles, make sure they use them. Clear up all the chemicals carefully when they're finished. Don't leave anything out in case someone swallows it.

Enamelling kilns. They get amazingly hot, so follow all the instructions about heat-resistant mats and advice to protect your child. They're suitable only for older children, and it's much better if you can see what they're up to while they're making the jewellery. Be sure you know what to do about burns (see page 145) as it's easy to catch a finger.

Kites. They can be fun, but keep them away from overhead power lines.

Babywalkers. I think babywalkers are toys because they just give baby a good time. They don't teach baby how to walk. Some babywalkers tip up or collapse. All babywalkers let baby scoot around fast so they can get at dangerous things – the fire, the oven, hot drinks, knives. Many babies have accidents with babywalkers every year. My advice is don't bother to buy one.

Good toys for different ages

Little children will get the most fun from toys they can easily use, and that help them find out about the world. Don't just buy a toy you want to play with. Most babies like the wrapping paper better than the present.

Good toys for different ages

	at birth to 6 months	at 6 to 12 months
What a child can do	wriggle and kick take notice of sounds watch something that moves start to reach out and hold things can't sit up without help	hold things pick things up and put them down put things in his mouth sit up (with help at first) lift up his head and chest when he's on his tummy
Good toys	rattles mobile for cot baby mirror on cot soft bricks and balls tie-on toys for pram and cot activity arches or mats NOT plush fabric toys	light things that are safe to chew like rattles and teething rings activity centres toys with a stick-on base so they can stick to a tray NOT babywalkers

	up to 3 years	up to 4 years
What a child can do	throw a ball screw a lid run and climb draw look at books match colours and pictures	pedal throw and catch copy, trace, cut with scissors string beads do easy jigsaws play pretend games
Good toys	trundle toys picture books picture dominoes chunky crayons balls water toys	tricycle scooter balls crayons, paper, scissors, glue beads to thread easy dressing-up toys dolls shops

Getting things changed

If you find something dangerous about a toy, complain. Get in touch with your local trading standards officer (TSO). They can check a toy for safety, get it taken off the market if it's dangerous, and prosecute the people who make it or import it. They can also alert other TSOs.

at 1 year to 18 months	*up to 2 years*
walk push and pull objects understand some words bang on the table	pick up small things hold a pencil build up bricks look at pictures copy grown-ups
push-along babywalker or trolley pull-along toys drum or xylophone plastic or rag books shape sorter (easy shapes) big cardboard boxes surprise toys like jacks-in-the box	paper and crayons banging toys like drums, hammers and pegs blocks picture books lift-out jigsaws toys that enable them to copy grown-ups like prams and cookers

up to 5 years	*at 5–8 years*	*at 9–12 years*
skip and hop plan and build count copy letters understand rules in a game		
playmats, toy cars, farms skipping rope picture books drawing and painting snakes and ladders and other easy games	dolls building and construction sets comics jigsaws ball board games imagination games fort and soldiers or spaceship with warriors	bicycle building and construction sets football electric cars and train sets plastic models skateboard dolls with different outfits computers

CHAPTER FIVE

Outings and Special Events

In this chapter I'll talk about outings you might take the children on. I'll also talk about other special events and fun things you might do at home.

Shopping

Shopping in the street

All babies and toddlers love going out in the buggy. Make sure you put your buggy up properly and that the safety-catches have clicked into place. Always use a proper harness, even if you're just going out for five minutes. Put the shopping on the tray – the buggy may tip up if you hang bags from the handlebars. If you take the dog, tie it to a post outside the shops and not to the pram or buggy.

The supermarket

It might not sound like an exciting outing to you or me, but it can be quite a treat for a baby or toddler. Supermarkets are built to make it easy to spend money. They're not built to help keep children safe.

Check the trolley. Some baby trays are at the back of the trolley or down the bottom. Some are high up on top of the shopping. If you have to put baby high up, it can make the whole trolley topple over. Don't use trolleys like this. Complain to the manager if necessary.

Check the toddler seat. You need to keep your toddler in. If the supermarket hasn't put a harness in, use your own safety harness to stop your toddler tumbling out.

Keep track of toddlers. There are a lot of things in supermarkets little ones can bang into or pull down on top of themselves. If your toddler likes to walk round with you, use a harness and reins so they stay close.

Watch their fingers. My baby once went round in the trolley eating a raw leek he had picked up. It's one thing to look silly. It's another if they pick up something small they can choke on, or something they can bite into dangerous pieces which could then choke them.

Crèches and Play Areas

When you go to shopping centres and supermarkets you often see a crèche or play area. Some are well run and have expensive, exciting toys your child does not have at home. Others have sloppy organisation. Here are some things to check:

♦ Sign in your child. Make sure they have a record of who your child is, and that only you are allowed to take them out. Make sure you have to sign out your child, otherwise they can't tell if children are still there, or if they've just wandered off.

♦ Make sure the children are kept in. There must be a barrier or a door – and someone waiting nearby – so they don't get out and get lost.

◆ See if big kids and little kids are kept apart. It's much safer if babies, little children and big children don't have to play together. Little ones can get hurt otherwise. Also, they don't want the same toys. A pull-along string toy that's great for a toddler can choke a crawling baby. A little bead can easily choke a toddler.

◆ If you're leaving a baby for some time, make sure there's somewhere safe where they can be put down to sleep. A proper crèche should have at least one cot. If it's a one-day crèche at a fête or a meeting, a carry-cot or play-pen will be quite adequate enough.

◆ Make sure there's somewhere to change baby. It needs to be safe, with room to get nappies, lotion and cotton wool out, without leaving baby. It needs to be clean, with somewhere to wash hands.

Shopping Centres

Shopping centres can be safe. At least you're away from cars and traffic once you're indoors. But there can be danger spots. There may be stairs and escalators where children can fall. Or fountains and ponds where children can drown. There might be railings and balconies where children can climb. You need to keep kids close to you so you can see what they're up to. A harness and reins are useful for toddlers. If you see something you're really worried about, complain to the management.

When you take the buggy, remember to strap your child in. Use the shopping tray for plastic bags of shopping. Don't carry them on the handle; this can easily make the buggy tip backwards.

Carry little kids on escalators (or look for a lift – it makes life much easier). Small children are slow to jump off at the ends. They can fall over, hurting themselves, or cause a pile-up.

Carry younger children on escalators.

Car parks

Be on the look-out for children. They may be running between cars. It's easy to reverse into a small child. Hold onto your children. Make them stay in the car until you get out and then let them out one by one. Teach them to stay close to you while you unpack the buggy and walk to the shops. Put the kids in the car before the shopping and the buggy.

Mother and baby rooms

It's nice to see that many more shops and shopping centres have mother and baby rooms and even parent and baby rooms to help fathers. Here are some things to watch for if you use one to change your baby:

♦ warming bottles. Bottle warmers are often full of hot water. If you've got a baby and a toddler, make sure the bottle warmers are pushed well back, with no flexes hanging down for children to grab

♦ changing surfaces. Make sure you can reach everything you need before you start to change your baby. Always stay close to your baby. It helps if there's a ridge to help keep baby secure, but don't rely on that. Stay next to baby until he's clean and dressed.

Swimming

Peter went out with his mates after school. They were all 10 now and they'd been able to swim for years. There was a local river they often went to. It was April and though the day was warm the water was surprisingly cold. After he'd swum out from the bank, Peter felt tired and couldn't swim back. He drowned before his friends could get help.

Water can be great fun. Children can splash and pour when they're little, swim and dive when they are older. But it's all too easy to learn to swim in a nice warm swimming pool, then have a big shock when you try to swim outdoors. Outdoor water can be cold, deep, with currents, gravel that slides under your feet or weeds that trap your legs. It's hard to swim in outdoor water – and dangerous. Swimming even in a warm pool is dangerous if the child is alone, tired or hungry (in case they get too weak to carry on) or has just eaten (in case they get sick and choke).

Rivers, canals and gravel pits

They're cold. They're full of problems:

Gravel pits:

♦ the edges will be sliding gravel that slips away as children try to get out

♦ children don't know how deep it is

♦ there may be weeds that make it look shallow when it's very deep.

Canals:

♦ the walls are man-made and steep

♦ locks are very deep indeed.

Rivers:

♦ the water is cold

♦ the currents can be powerful

♦ children don't know how deep it is.

Reservoirs:

♦ the water is cold

♦ it can be shallow, then suddenly get deep.

Teach children not to swim in places like this.

Diving into the water

If a child (or anyone for that matter) dives into shallow water and hits their head, they can break their neck and end up paralysed. The best place to dive is in a special diving pool at the swimming baths, where it is deep enough to be safe. It's not safe to dive into an ordinary pool at home or on holiday. Children should go in with feet first the first time in any water.

Falling in the water

If someone falls in, don't go in after them except as a last resort.

◆ **Reach** out to the child. Use anything to help you reach out further, like a branch, a towel, or belt.

◆ **Throw** out anything that will float on the water. It could be a rubber ring, a beach ball, even a plastic bottle.

◆ **Wade** out into the water. If there's a current, it's safest to use a human chain, to hold onto the shore.

◆ **Row** or paddle out to the child if there's a boat.

◆ Go out and swim back with the child only if you're a strong swimmer and you've learnt to do life saving. Otherwise, you yourself may drown.

◆ Go for help instead. Call 999 for the police or lifeguard.

Drowning

If someone is drowning, carry out the ABC of resuscitation (see page 135). Don't give up before you get to hospital.

At the seaside

◆ Know where young children are. Keep an eye on them. Be extra careful if they run into the sea.

◆ Check the beach for sharp glass or old needles. Get your children to wear sandals or flip flops.

◆ Check the sand for dog's mess.

◆ Keep to the swimming tips (see page 110). Don't let anyone swim far out to sea.

The swimming baths

Swimming baths are a good safe place to swim. There are people about who know what to do. People who have thought about safety and stick to safety rules.

Babies and toddlers

Babies and toddlers can enjoy themselves at the swimming pool splashing about in nice warm water, and even learning to float. But remember, they will forget everything they know if they fall into a cold, deep river. For parent and baby swimming:

♦ Make sure the water is warm. Little children get cold quickly.

♦ Don't keep them in the water for more than 20 minutes.

♦ Don't let them go head under. They can easily swallow a lot of water and might get too much water in their bloodstream.

♦ Make sure children wear costumes. That way, if someone does a dirty bottom, the mess is held nearby and doesn't float out into the pool.

A toddler who's learnt not to be afraid of water can be careless near pools and ponds, so watch them extra carefully.

Older children

No one under eight should go swimming without a grown-up. Teach children not to run about wildly, in case they slip, or knock down other people. They mustn't jump in the pool on top of other swimmers. They must dive only in special deep pools (these might be separate) or the deep end. Check the depth before they dive.

A good age to learn to swim is about five or six. That's old enough to listen to a teacher and learn fast, and young enough to get to be an expert swimmer, or just have fun later on.

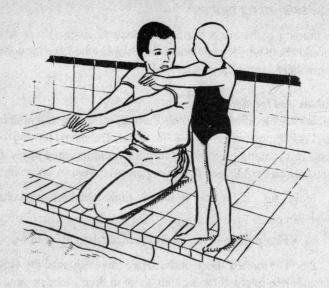

It's good to learn to swim in the pool, but swimming out of doors is another matter and can be dangerous.

Is your pool safe?
Here are some ideas for an extra safe swimming pool. If you're worried about a local pool, complain.

♦ The changing rooms are away from the pool side, so kids can't rush straight into the water.

♦ There's a non-slip surface (it may be just textured tiles) round the pool.

♦ There's a special shallow paddling pool for very young children.

♦ There's a special deep pool for diving. Divers at the deep end can hit swimmers.

114

◆ There are warning signs telling you not to dive where the water's shallow.

◆ There are other warning signs about rushing about, or jumping in.

◆ There are lifeguards on duty.

◆ The lifeguards can see what's going on – they have special chairs, high up, so they can watch the pool.

◆ Any parts closed for repair, or because there's no lifeguard, are shut off, so children can't get there.

◆ Any empty pools are closed off. Jumping into an empty pool can be dangerous. So can diving into a small amount of water at the bottom.

◆ Open-air pools are fenced off so children can't get into them when they're shut.

Fun pools

A lot of councils now have pools with water slides and chutes. These can be terrific fun, but kids have to be sensible when they use them.

◆ Don't let children go down too close to one another.

◆ Don't let children wait inside for their friends to catch up with them. Sliding down in twos or threes makes for extra accidents.

◆ **If someone hurts themselves** on the slide, make sure the pool staff know. This is how mistakes in designing water chutes are found out. At one pool, 63 people were cut near the right eyebrow after sliding down, but they didn't tell the staff. The staff only found out because someone at the hospital got fed up putting stitches in people's eyebrows and told them. A bend was made less sharp and the accidents stopped.

Sports and games

Most sports don't lead to bad accidents. Still, children shouldn't be allowed to go berserk. They should wear sensible shoes if they're kicking a ball. If there are special pads or protectors for the sport, children should wear them. If your child plays an unusual sport, get advice from the association in charge. The Sports Council (see page 155) will let you know what it's called and how to get hold of it. Whatever sport your child takes up, don't push them too hard. They won't have fun and are more likely to get hurt, strains, or arthritis later on.

Ice skating

Most accidents at the rink aren't serious.

♦ Make sure children wear sensible clothes. Beginners fall a lot and thick trousers or jeans and a jumper will help cushion them. Glamorous outfits are for those who know how to skate.

♦ Let your children have some lessons. People who can skate properly don't get hurt nearly so often.

♦ If there are special sessions for good fast skaters, keep beginners away.

♦ Check that your local rink has proper first aid arrangements for helping people who've been hurt. If they haven't, complain. And keep your child away.

Rugby

Most rugby accidents are just bangs and knocks. There are sometimes serious accidents when young people get broken necks. Some parents don't allow children to play rugby at all. If your child is a rugby player, they must get proper training from a coach who knows what they're doing.

Horse riding

Most accidents happen when teenage girls fall off horses. People also get kicked or horses fall onto children. The worst accidents are blows to the head.

♦ Make sure that the horse and rider go together – a solid steady horse for beginners.

♦ Make riders wear sensible shoes and clothes.

♦ All riders under 14 must wear a proper safety riding helmet (to BS6473 or BS4472). That's the law. Check the hat fits and that it's done up properly.

♦ Make sure child riders learn how to ride on the road safely through the British Horse Riding Society and Road Safety Scheme.

♦ Drive sensibly near horses. Don't hoot at them, or rev up. Overtake them calmly and slowly.

Fun-fairs and theme parks

Every year, about 500 000 000 rides are taken at fun-fairs and on roundabouts. The biggest theme parks have about 2 000 000 vistors a year. It's unusual for anyone to be badly hurt. That's because there are rules and regulations about safety that all fun-fair owners have to obey. The equipment has to be checked several times a season. So frightening accidents like the carriage flying off the roller coaster are highly unlikely. There are ways you can help your family be safe from minor bumps and bangs:

Take plenty of grown-ups. It's hard to have eyes in the back of your head. On some rides you'll need an adult to help and stay with every child. This is especially true for water rides, even if they move slowly.

Read the signs and labels. If the rides aren't suitable for small young children – or for big tough heavy ones – don't let these children go on them. It's useful if you know how tall your child is, as many rides are only for children above or below a certain size, not a certain age. That's because they need to hold on to the handles or be wedged in by bars, and only people of the right size will fit.

Keep alcohol out of theme parks. It can only help make people silly and dangerous.

Check there are proper fences round rides and pools. If there aren't, complain.

For advice on car parks see page 109, and for changing areas see page 110.

For extra enjoyment, do look at the guide. There are often special play areas for younger children, with safe rides and other treats like ball pools. This is better for them than long queues for daring rides they won't be allowed to go on – even if it isn't such fun for grown-ups.

Farms

Farms can be dangerous places for children. That's true even for country children who live there all the time. They're even more dangerous for town children on holiday who don't know what to expect and are excited.

◆   Keep children away from machinery. Even if it's turned off, kids can pull over heavy things and be crushed. Rides on tractors can be dangerous too, and shouldn't be allowed.

◆   Keep children away from animals, except with a grown-up who knows the animals well. Even a minor accident, like a cow treading on a foot, can do a lot of damage.

♦ Make sure children don't play in pools. They're cold and can have slippery edges. Children can also drown in water butts and drinking troughs.

♦ Beware of slurry pits. They look solid, but they're just loose mud. Find out where they are and keep well away.

♦ Keep clear of stores. Farmers may have chemicals like weedkillers or rat poisons which could poison your child.

Fishing

It's easy for kids to fall in when they're fishing. Teach children these safety tips:

♦ Never go fishing alone. There will be no one to help if you fall in.

♦ Keep away from steep, slippery or crumbling edges.

♦ Tell someone where you are going and roughly when you expect to be back.

♦ Don't go if the weather is bad (listen to the weather forecast before you go).

♦ Be careful with hooks and flies. They can slash someone's face or eye.

Holidays and travel

I've already given tips about being safe in the car (see pages 76–82). Most journeys on coaches and trains are very safe. So are plane trips, even though every crash makes big headlines. If you're taking your baby on a plane, you may be allowed to use their baby car seat. Check with the airline for up-to-date advice. That way, you can sit them as safely as possible. To look after children, you need to keep an eye on them. Don't let yourself get too tired. Take a rest if you're driving.

Be extra careful:

◆ if there's water. Watch children on gang planks and decks of ferries or ships

◆ if there's hot food. It's easy to spill a hot drink or a hot meal and scald a toddler in a café or a restaurant just like you could at home

◆ if there's traffic. Don't get muddled by cars driving on the opposite side of the road. Remember drivers may be more aggressive in some countries than they are in Britain.

In your hotel

There are nasty fire accidents in hotels abroad every year, because fire safety rules aren't as strict as in the UK, and people don't stick to them. You should take special care to know how to escape.

◆ Check you know how to get out of your room if there's a fire.

◆ Make sure fire escapes aren't blocked.

◆ Make sure fire doors aren't wedged open.

◆ If fire breaks out, check the door. If it feels hot, leave it closed. Use sheets to block off gaps around the door to keep smoke out. Try to escape through a window. If the door's cold, go out through it. Crawl out towards the fire escape keeping low on the ground in case you come near smoke. Shut the door behind you to help stop fire spreading.

◆ If you think there are fire hazards, complain to the hotel and to the tour operator.

WATCH OUT! If hotel pools aren't fenced off, children can wander in and drown.

Hotel swimming pools

Is the pool fenced off? Is there a children's pool? Is there any supervision? Does the pool suddenly get deep? Is it shallow without any warning, so you might dive in and be injured? If you find problems, don't be afraid to complain.

Parties

Parties for small children

It's hard enough to keep an eye on one or two children, and it's impossible to watch over 15 or 20. So if you're having a house full of youngsters, think safety before they arrive.

121

◆ Check over your house for dangers like poisons and matches. They shouldn't be anywhere a child can reach.

◆ Lock rooms you don't want them to use.

◆ Make sure the front door's shut. Only open it to let kids in and out while a grown-up is there.

◆ Have a carton for presents and wrapping paper. Don't let kids be hurt on the extra clutter if it gets on the floor or stairs.

◆ Keep children out of the kitchen. There's bound to be much too much going on for you to watch them.

◆ Let them play in the garden only if there's a grown-up to watch.

◆ Be careful with the birthday cake matches. Keep them out of sight until you need them and put them away as soon as you've used them.

Bonfire parties

We've all heard such a lot about dangerous fireworks that even children are usually careful with them. But because they can do so much harm, we must still 'think safety' every bonfire night. It's safest to go to a properly organised display.

Always obey the firework code if you're having a firework party at home:

◆ Keep fireworks in a closed box.

◆ Don't put fireworks in pockets.

◆ Don't throw fireworks.

◆ Never go back to a firework you've lit, even if you think it's gone out.

◆ Keep a bucket of water handy in case the flames get out of control.

◆ Keep your bonfire away from fences and sheds which might catch fire.

◆ Keep away from the bonfire. Children and grown-ups' clothes can easily be set on fire, or they may tread on something that's burning or smouldering.

◆ Make sure the bonfire stays small. Don't use petrol or paraffin to help it burn. Flames can spring up and might flash back into the container. Don't light it at all if there is a strong wind blowing.

◆ Be careful with matches. Don't leave them lying about in case children play with them.

◆ Ban booze at bonfire parties. Most firework accidents happen because people get careless and reckless. Alcohol doesn't help matters.

◆ Don't let young children touch dead sparklers; they stay hot for a long time.

◆ Make sure you know what to do about burns (see page 145).

Christmas safety

Most accidents to children happen when people have many other things to think of and they're under a lot of pressure. To me that sounds like Christmas for most people. The house is full of people, everyone is excited or bad-tempered, and there's tons of extra cooking and work. You need to keep on 'thinking safety'.

♦   Be careful in the kitchen. It's easy to be so rushed that we forget how dangerous pans of vegetables, steamers and hot ovens can be. Keep being safe like you usually are.

♦   Don't get burnt down. All the cooking, and candles or faulty fairy lights on trees, can easily start a fire. Wrapping paper from presents and decorations burn easily. Take care with matches. Put a smoke detector on your Christmas shopping list.

♦   Don't trip and get hurt. Keep the clutter under control. Grandmas can break a hip if they fall over new toys. Old cartons for scooping everything up can be handy.

♦   Keep booze away from the kids. They can come to serious harm if they come down in the evening after a party and finish off the leftovers in glasses or chew up a few cigarette ends.

♦   Keep peanuts away from kids. They're a nice party nibble, but not for children (see page 21).

Getting things changed

You can make life safer. You can be extra careful with your own kids, and with other people's kids in places like car parks. You can also complain – to swimming pool staff, to supermarket managers, to travel companies. Don't think no one will listen. You are the customer and you are important. If the local people won't listen, try head office. Try a trade organisation like the Association of British Travel Agents. Try the Citizens Advice Bureau for other ideas about where to complain.

CHAPTER SIX

Your Child's Carers

I've talked a lot about how your house, the things you buy or the way the streets around your house are laid out can make life safe or not so safe. Sometimes safety comes down to people – the people in your family or people outside your home who look after your children. In this chapter, I'll talk about how they can be safe, and how they can make life dangerous.

Teaching your child to be safe

Older children are able to learn how to be safe for themselves. There are things you teach them every day, when you tell them not to touch or to watch for cars. Sooner or later they need to go it alone. They need to know how to spot dangers and why safety lessons are important. They need to know what to do if things go wrong. Here are some ideas about games you can play that will help them to think about safety for themselves.

What is a danger?

What worries your kids? What are they scared of? What do they want to keep safe from? Your kids won't sit down and talk to you about it, but you could start to find out by drawing a big picture.

You will need a large piece of paper and some crayons or coloured pencils. Ask your child to draw a picture of themselves in the middle of the page and put a circle around their drawing. Tell them they are in the circle to keep safe from dangerous things. Ask them to draw pictures around the edge of the circle of these dangers (see illustration on page 129). Then ask them to talk about the picture with you. Even four- and five-year-old children can play this game. You'll find out a great deal about how kids see the world. They often have funny ideas. Little ones are often more worried about monsters than the traffic on the road. It's also a chance to help them keep safe from the things that worry them. You can tell them safety tips too.

Kids have to learn to spot hazards for themselves if they're going to keep safe. You could use the illustrations in this book to help you help them to spot the dangers. Ask your child what the people in the pictures are doing, what could be dangerous or ask them how they are keeping safe. Sitting down with a book can help too. It makes your child concentrate instead of rushing about.

Hazards in the home

Ask your child to accompany you on a tour of the house. Where are the hazard areas? What does the family need to do? It can be fun if you make signs that say DANGER SPOT in big red letters for your child to place around the home. This way they can tackle the dangers with you and share in keeping the little ones safe.

Learning to concentrate

Many accidents happen because kids don't concentrate. They rush about and fall over their own feet. They dash into the road.

Ask your child to draw dangers round the picture of themselves and discuss them with you.

They gobble food and choke. So concentrating really helps kids to be safe. See if you can get them to settle down for a few minutes to play a concentrating game. Ask them to sit on the floor and close their eyes. Tell them they're going to practise concentrating. Ask them to listen out for noises outside the room for two minutes. Then ask them to listen out for noises inside the room. Make a special quiet noise like crunching a paper bag. Tell them to open their eyes and tell you what they could hear. Ask them if they can understand how concentrating helps to keep safe.

Preparing for accidents

Being prepared and practising can help you tackle real problems when they arise. We've already talked about the family fire drill (see page 48), but older kids can also play a telephone game. You will need a toy or an old phone. Or you can unplug the real phone if you've got the right sockets. Tell your child to pretend that an accident has just happened: grandma has fallen down and she can't speak, you're on holiday and you see someone drowning, the chip pan catches fire, you see a burglar climbing into someone's window.

Do they know what to do?

Can they practise calling 999?

Can they give their name and address correctly?

Clubs and Cubs

Many organisations run groups for kids. They often teach children a lot about safety, first aid and how to be sensible and caring people. The Boy Scouts, Girl Guides, St John Ambulance and the Red Cross have groups for young children, including some for six-year-old children and up.

Your child's school

Does you child's school do anything to help develop your child's sense of safety? Ask your head teacher or parent governors. The school could arrange visits by the fire brigade or the police, or local road safety officers and home safety officers. It could use teaching packs about safety as part of the curriculum, or join a young citizens programme which helps children practise being safe. It could even run a first aid course for the older pupils.

The baby-sitter

Everyone needs to go out. If you have one or more children, you deserve to even more. Usually it's easier to get a baby-sitter to come to your home.

Know your baby-sitter. It's best to have someone you know or have heard about from a friend. If you have to find a baby-sitter from an advert in a newsagents or the local paper, get references. Phone up people they've worked for and check them out.

Choose someone sensible. If your baby-sitter is under 16, you could be responsible for them as well as your own kids if something goes wrong. Remember, they're going to be alone in your house. Don't have anyone who'd make you worry.

Have a rule about boyfriends and girlfriends. You want the baby-sitter to be listening out for your kids, not having her own party. If you don't want boyfriends round, let them know from the start.

Explain everything carefully. Tell them about safety items like the smoke detector, and how to unlock the windows as well as how to work the TV.

Have a rule about smoking. Cigarettes can start fires, so make sure your baby-sitter is safe with matches and cigarettes, if you let them smoke.

Leave some emergency numbers for:

♦ where you are

♦ a friend or a relative who can come round

♦ the doctor.

Make sure your own address and phone number is on the phone so if they have to call the fire brigade in a panic, they will know where they are.

Help your baby-sitter to be safe too. Don't expect them to go home late by bus or tube. Run them home or at least give them the taxi fare.

Working mothers

I'm a working mother and I'm lucky. I can leave my two with a trained nanny. They're in my home, which I've checked for safety. My nanny has learnt about kids, about safety and first aid.

There are many other people who care for kids while Mum's at work. They can be good too – caring people who help kids grow and learn. Your town hall may have an 'under-eights adviser' or someone else with the same sort of job; these people can give you advice about child-minders and other carers in your area.

A nursery, crèche or playgroup

Most nurseries have some trained staff. They know how to look after kids. They may have done first aid. The laws about safety are basic and there is much more a nursery can do. If you're choosing a nursery, it's important to have a look round and take time to talk to the staff. That way, you'll find out if you feel comfortable with them. Can the nursery be a home away from home for your children?

Here are some ideas about things you can check:

Atmosphere	Do you like the atmosphere? It doesn't matter if it's quiet or noisy, as long as the staff are in control, and you feel comfortable.
	Is there a routine so that the children know what's going on?
	Are the activities planned and organised or do children just muck about?
Staff	How many of the staff are trained?
	How many of the staff have done a first aid course?

Organisation	Are there proper rules about leaving children and picking them up?
	Is there an accident book?
	Is there a first aid box?
	Is there a fire drill and a fire practice?
Layout	Are the kitchens separate? Can they keep children out?
	Is there a special place for children to eat, so that kids who are playing don't knock into hot food and drink?
	Are there steps where children may fall?
	Can children get out the front into a road?
	Is the garden or outside play area fenced off so children can't get lost?
	Is there water like a paddling pool or pond where children could drown?
	Is there somewhere to store toys, so there aren't piles of clutter on the floor?
Equipment	Is there somewhere soft to fall under slides or climbing frames?
	Are toys (indoors and outside) in good condition?

If your child gets hurt at a nursery, or you see something that worries you, do complain.

The child-minder

It's much better to choose a registered child-minder. Their home will have been checked by social services to make sure it is safe. Often they will have done a special course on how to look after children.

When you're choosing a child-minder, talk about keeping your child safe. Ask them about fires, scalds, cuts and poisons, and many of the safety ideas we've talked about. Make sure they've got some safety equipment like fire-guards and safety barriers. A smoke detector is a good idea too. If they're going to use the car, make sure they have seats for the kids. If they mind several children of different ages, see that they have different things for them to do. Ask if they've done a first aid course.

A good child-minder will care about your kids. She'll want to help keep them safe. She'll be happy to talk about safety to you. You have to feel comfortable with her and with her home. She's got to have the same feelings about safety as you. That's important for you, so you can relax when you leave your children in her home.

A friend or relative

It's nice to be able to leave your child with someone you know well. If you 'think safety' with your child, ask them to 'think safety' too. Go over all the safety tips. Make sure their home is safe like yours. Don't be shy.

Granny's home and older people

Most kids love to visit granny and grandad. They give lots of cuddles and treats. Your parents may be young and lively, and have the same ideas about safety as you. But they may be older and have safety problems in their homes.

Things were different in the old days. There weren't a lot of fast cars whizzing past people's homes. It was all right to play in the street. We have to be much more careful now than we used to be.

Older people have more medicine. They often have tablets and pills. They can't take the tops off child-resistant bottles if they have arthritic fingers. They like to keep pills handy in bedside drawers and in handbags.

Older people have clutter. We all collect ornaments and bits and pieces as we go along. By the time we get old, we have all sorts of things. They can fall down, they can squash children, they can break. They might be little bits a toddler can swallow and choke on.

Older people can be careless with cigarettes. They might smoke in bed, and could drop off to sleep with a cigarette. A fire could easily start.

Talk about safety to Granny and Grandad. Help them to change things to keep the little ones safe. Buy grandparents a smoke detector. Older people who can't get about quickly need more time to escape if there is a fire, just like toddlers who can't climb and run.

CHAPTER SEVEN

If an Accident Happens

This book is about stopping children from being hurt. It tells you how to make sure nothing really serious happens to your kids. But we can't stop every bump or scratch. We can't wrap kids in cotton wool. They've got to have some space to grow and to learn for themselves.

It's not easy to tell you much about first aid on paper. It's better to take a special first aid course if you can. Get in touch with St John Ambulance or the Red Cross for extra help. If you learn something about first aid, it will be easier to keep calm and to do the right thing in an emergency.

General tips

I split my head open. I was climbing up and fell back and hurt myself. I got a bandage. We phoned Auntie Doreen. She put the bandage on. I wasn't crying. I got jabbed.

♦ Don't panic – things often seem worse than they are. For example, even small cuts on the head bleed a lot. If you panic you will only frighten the child more. So keep calm.

♦ Don't get injured yourself. If there's a gas leak, turn the gas off and open the windows. If it's electrocution, turn off the power and pull the child away from the electricity with a broom. If it's a road accident, bar the traffic before you run into trouble.

♦ Don't try to do too much. Get medical help if you're in doubt.

Very serious injuries

Unconsciousness and the ABC of resuscitation

Resuscitation is what you need to do if someone is unconscious and they're not breathing for themselves. Perhaps their heart has stopped beating too. Resuscitation is for people who are seriously hurt.

Resuscitation has an ABC to help you remember what to do:

♦ A (Airway)

♦ B (Breathing)

♦ C (Circulation).

I'll explain what this means:

A (Airway). That means a way for air to get in and out of the lungs. Clearing the airway means making sure there's nothing to block air moving in and out. It means taking anything off the face and out of the nose, and clearing the mouth out quickly of blood or vomit. Remember: don't spend ages poking inside someone's mouth. You may do more harm than good. You may push something down.

B (Breathing). This means breathing for someone if they're not breathing on their own. You put your mouth over theirs, or nose and mouth for a baby or small child, and breathe into their lungs. Be gentle if you are trying to breathe for a young child. You can damage their lungs if you blow too hard.

C (Circulation). This is the blood being pushed around the body. If someone's heart stops, you take over the pumping for them by pressing on their chest.

So, if you find a child unconscious, follow the ABC of resuscitation:

1   Check the response. Shake them gently and shout wake up.

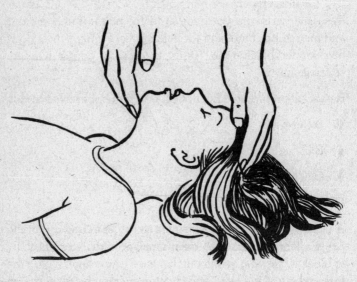

Put one hand on the forehead and the other under the chin to open the airway.

2   Open the airway. Lay them on their back on a hard surface. Take anything off the face. Put one hand on the forehead and the other under the chin. Gently tilt the head back. Look in the mouth and clear any obvious blockage with a single wipe of the finger.

3   Check if they are breathing. Look for the chest going up and down and listen and feel for breath. If they are breathing, place them into the recovery position (see page 139).

4   Start artificial ventilation. Pinch their nostrils together and seal your lips around their lips; for a baby, seal your lips around their nose and mouth. Blow gently into their lungs. Watch for the chest to rise. Do not blow too hard. Remove your lips from theirs and let the air come out. After two breaths, check for a heart beat.

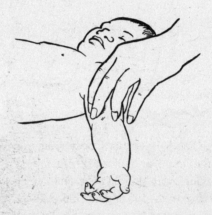

Feel for a pulse, on the arm, if you are treating a baby.

5   Check that the heart is beating. Feel for a big pulse – in the neck or on the arm if you are treating a baby. Use your fingers, not your thumb as you feel your own pulse if you use your thumb. If you do feel a pulse, continue to breathe for the casualty and check the pulse at regular intervals. If you do not feel a pulse, start chest compressions.

6   Chest compressions. Find the place where the bottom of the ribcage meets the breastbone. For an adult, place the

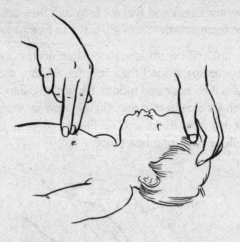

Chest compressions.

heel of one hand above this. Cover your hand with the heel
of your other hand, and interlock your fingers. Press down,
keeping your arms straight. For children, use one hand
only, and for babies, two fingers only. Release pressure. For
adults, press fifteen times, and then blow into the lungs
twice. For children and babies, press five times, and blow
once. If two people are present, one person can press the
chest and stop every five presses for the other person to do a
breath. Check the pulse regularly. Once the heart starts
beating, carry on with artificial ventilation until breathing
starts and then move on to the recovery position. For a
baby, find the line between the nipples. Then find the spot
in the middle, one finger's breadth down. Press here with
two fingers. Press down for half an inch to one inch about
100 times a minute. Do five pushes followed by a gentle
breath like you would for an older child.

The recovery position.

7   The recovery position. This is the position for someone who is unconscious but breathing on their own. The airway stays open and any vomit or other fluid will flow from the casualty's mouth. Turn your casualty into the recovery position shown above. Remember to protect their face, especially when turning them.

People who are badly hurt and unconscious

Don't try to move a child who is seriously hurt unless there is an emergency like a fire. This is important if you think there are broken bones, a head or spinal injury. Turn the patient into the Recovery Position, as explained in the ABC of Resuscitation, and send for an ambulance.

Get extra help

Don't leave an unconscious child alone in case they start to choke. Call for an ambulance, or get someone else to call. Try signalling for help through a window if necessary.

Bad injuries

Bleeding

Press firmly on the place that's bleeding. Use a clean cloth if you can find one quickly. A tea towel or hanky will do fine. Otherwise use your fingers. Lifting up an arm or a leg can help slow bleeding. Don't do this if the arm or leg might be broken. Keep pressing until the bleeding stops. This can easily take 10 minutes. If one pad gets soaked through, don't take it off, just put another pad on top. Cuts on the scalp always bleed a lot, so be patient.

Go to your local Accident and Emergency department for:

◆ a big cut that might need stitches

◆ a cut where the edges don't come together

◆ a cut over a joint where the edges keep coming apart

◆ a cut on the wrist which may have hurt the nerves, or tendons that move the fingers

◆ a dirty cut, or a big cut that happens outdoors, which might need a tetanus injection or an antibiotic

◆ a cut with something left inside, like a piece of glass. You need to press round it, not on top, to stop the bleeding. Do not attempt to remove the object. Leave the object inside the cut and go to your Accident and Emergency department immediately.

Not all big cuts need stitches. Children heal quickly, so cuts can often be held together with special sticky strips.

Don't give a child anything to eat or drink if you think they may need an anaesthetic. Remember, it can be easier to put a child to sleep for some stitching jobs, even if an adult could manage with a local anaesthetic.

Broken bones

A bone may be broken if an arm or leg is bent in a funny way, or at the wrong place. The area over a broken bone may swell up and look red. It also hurts – a lot. If your child has a bang and they have broken something, they won't let you touch it to check! They won't want to move a broken arm or leg. Don't move a child with a broken bone unless you have to. You might be able to move a child with a broken arm. Use a sling (made out of any material or even a belt) to give support. If bits of bone are sticking out of a wound, cover the area with a clean cloth. Make the child comfortable, and either go to the local Accident and Emergency department or call an ambulance.

Don't give a child anything to eat or drink. They may need a short anaesthetic while broken bones are straightened out.

Head injuries

I had stitches in my head because I banged it. They sewed the hole back up.

Head injuries can have serious consequences because the brain controls the rest of the body. That means the breathing and the heartbeat too. If a child has a head injury, try not to move them, you may do more harm than good. Just pull the jaw forward gently to stop the tongue blocking the air passages.

141

Call an ambulance, even if the child has only banged or cut their head if:

♦ they have been unconscious even for a short time

♦ they have a large cut on the head

♦ there is blood coming out of their ears or nose

♦ they have a fit.

Keep a close eye on the child for at least 24 hours. Sometimes people seem to be fine immediately after banging their head. But slow bleeding may start inside the skull. The blood builds up and presses on the brain and can cause a lot of damage. So always send for an ambulance if a child is sick, gets drowsier than usual, or has a bad headache a few hours after a head injury.

Choking

This is what happens when something gets into the windpipe, blocking air to the lungs. When we choke, we cough and splutter. This usually shifts the blockage. Even a baby can cough and recover from choking on its own.

If a child has a choking fit, coughs and then seems better, don't worry any more. You don't need to see a doctor if they seem back to normal.

If a child has a choking fit, coughs and splutters and can breathe again, but keeps coughing, or has odd noisy breathing or a funny voice, go to the Accident and Emergency department straight away. Don't try to shift the blockage. You may make things worse.

If a child chokes, and goes blue and can't breathe or speak, you need to be quick to shift the blockage and save their life.

For a baby or a young child:

♦ Don't waste time poking in its mouth. You may push the blockage further down.

♦ Lay the baby/child with its head downwards on your lap, supporting their head and shoulders. Slap smartly between the shoulders for up to four times.

For an older child or adult:

♦ Again, don't waste time poking in their mouth. You may push the blockage further down.

♦ Bend them over or place over your knee and slap them between the shoulder blades. Do this up to four times.

For all ages:

♦ If they stop breathing, follow the ABC of resuscitation (see page 135). Call for an ambulance.

Suffocation

This occurs when the air passages are blocked by something over the face and nose.

♦ Clear away the obstruction.

♦ If the child is not breathing, follow the ABC of resuscitation (see page 135).

Strangulation

This is when something round the neck stops someone breathing properly.

♦ Unwind the string or cord.

◆ If the child is not breathing, follow the ABC of resuscitation (see page 135).

Poisoning

It is frightening if you think your child has swallowed pills or chemicals. Remember, most things children swallow don't do them much harm.

For pills:

◆ Check: did the child really take them? Look on the floor and under the table. It may be a false alarm.

◆ Look for the bottle or a sample pill to take with you.

◆ Call for an ambulance immediately.

◆ Don't give your child salt and water to make them sick. This is more poisonous than a lot of pills or medicines.

For household chemicals:

◆ Do not give them anything to drink. Moisten the lips if they are burning.

◆ Look for the bottle or jar to take with you.

◆ Call for an ambulance immediately.

◆ Don't try to make your child sick. Salt and water is poisonous. Some household chemicals can hurt the lungs and if a child is sick, something harmful can go the wrong way.

◆ If your child becomes unconscious, or has trouble breathing, remember your ABC of resuscitation.

144

Burns and scalds

Where the cold water went, it all got better. It's only the bits I missed where she's got a mark now.

Immediately put the burnt part under a running tap or into a bath or bucket of cold water. The cold water takes the heat out of the burn. It can make the difference between a deep burn that takes the skin off or a tiny pink mark. Don't waste time taking off any clothes. Leave the burnt part under the water for at least 10 minutes. This seems like a long time but you must carry on. Continue until the pain stops.

♦ Take off anything tight like jewellery or belts – burns can easily swell up.

♦ If there is a big burn or scald, keep it clean. Wrap it up in a clean cloth, use clingfilm or put it inside a plastic bag, both from a new roll. Go to your accident and emergency department.

♦ Don't rub on butter or ointment. It will all have to be cleaned away and this hurts a lot.

♦ Don't prick blisters. You will let in germs.

Minor injuries

Things up noses or down ears

If your child puts a bead or bit of toy up their nose or down their ear, get help from a doctor. It's hard to pull these things out. They just end up going further in if you poke about.

Small cuts and bruises

Press on cuts to stop bleeding. Clean them with water. Wash with soap if they're very dirty. Cover with a plaster or small dressing if necessary.

Eyes

If you can see what is troubling the patient, such as grit or an eyelash, try to remove it very carefully. Encourage the patient not to rub the eye. If you cannot see anything, or it is difficult to remove, cover the closed eye with a clean pad, such as a hanky, and take them to hospital.

If a child gets a splash of something in the eye, hold them or help them lean back under a tap. Run clean water into the eye for 10 minutes. Then go to hospital.

Bites and Stings

Dog bites

Wash and clean the wound. Get advice from a doctor.

Insect stings

Pull out the sting with tweezers if it's been left behind. Bicarbonate of soda in water is useful on bee stings. You can use calamine lotion to soothe any redness from insect bites and stings. If a child gets a sting inside the mouth, or they get a sting and it starts to swell up fast, call an ambulance immediately.

Family medical kit

Here are some useful things to keep together somewhere handy. Most are for first aid, but there are some ideas for medicines too. Keep the kit somewhere clean and dry (not in the bathroom), and out of the reach of children.

◆ Assorted sticking plasters.

◆ Assorted sterile gauze dressings (dressings that don't stick to wounds can be especially useful. Ask your chemist. They don't hurt when you take them off grazes).

♦ Roller bandages and triangular bandages.

♦ A crêpe bandage for sprains.

♦ Soap (to wash dirty wounds).

♦ Safety pins.

♦ Scissors.

♦ Thermometer or temperature strip that you can put on a baby's forehead.

♦ Tweezers.

♦ Paracetamol (the right sort for your child's age).

♦ Antiseptic cleansing wipes.

♦ Calamine lotion.

♦ Any other item recommended by your GP.

Useful Publications

Children and roads: a safer way
Department of Transport
London 1990
From Road Safety Division of Department of Transport

Design considerations for pelican and zebra crossings
Department of Transport
Advice note TA 52/87
From DOE/DTp Publications Unit, Building 1, Ruislip, Middlesex, HA4 0NZ

First Aid manual
The authorised manual of St John Ambulance, St Andrew's Ambulance Association and the British Red Cross Society
Dorling Kindersley
London 1987
From bookshops

Friends of the Earth guide to traffic calming in residential areas
Friends of the Earth
London 1989
From Friends of the Earth

Highway code for young road users
County Road Safety Officers 1989
From your local road safety officer

Keep them safe
Child Accident Prevention Trust
London 1990
From health visitors and health promotion units

Keep your baby safe
Child Accident Prevention Trust
London 1990
From antenatal clinics and health promotion units

Measures to control traffic for the benefit of residents, pedestrians and cyclists
Department of Transport
Advisory leaflet 1/87
From the Road Safety Division at the Department of Transport

Pedestrians, an action guide to your neighbourhood problems
National Consumer Council
London 1987
From National Consumer Council

Playground management and safety
National Playing Fields Association
London 1989
From the NPFA

Playground planning for local communities
National Playing Fields Association
London 1989
From the NPFA

Pro-bike
Friends of the Earth
From Friends of the Earth

Useful Addresses

Association for Consumer Research
2 Marylebone Road, London NW1 4DX
Tel 071 486 5544
(Publish *Which* magazine)

British Horse Society
British Equestrian Centre
Stoneleigh, Kenilworth, Warwickshire CV8 2LR
Tel 0203 696697

British Red Cross Society
9 Grosvenor Crescent, London SW1X 7EJ
Tel 071 235 5454
(Your local branch will be in the telephone book)

British Standards Institution
Linford Wood, Milton Keynes MK14 6LE
Tel 0908 221166

Child Accident Prevention Trust
28 Portland Place, London W1N 4DE
Tel 071 636 2545

Department of Trade and Industry
Consumer Safety Unit, 10–18 Victoria Street, London
SW1H 0NN
Tel 071 215 3215

Department of Transport
2 Marsham Street, London SW1P 3EB
Tel 071 276 3000

Fair Play for Children in Scotland
Unit 29, 6 Harmony Row, Govan, Glasgow G51 3BA
Tel 041 425 1140

Fire Prevention Officer
At your local fire brigade – look up 'Fire' in the phone book and
ring the enquiries number

Friends of the Earth
26–28 Underwood Street, London N1 7JQ
Tel 071 490 1555

Girl Guides' Association
17 Buckingham Palace Road, London SW1W 0PT
Tel 071 834 6242

Home Office
Queen Anne's Gate, London SW1H 9AT
Tel 071 213 3000

Home Safety Officer
At your council offices

National Children's Play and Recreation Unit
359–361 Euston Road, London NW1 3AL
Tel 071 383 5455

National Consumer Council
20 Grosvenor Gardens, London SW1W 0DH
Tel 071 730 3469

National Playing Fields Association
25 Ovington Square, London SW3 1LQ
Tel 071 584 6445

Play Wales
10/11 Raleigh Walk, Atlantic Wharf, Cardiff CF1 5LN

Playing Fields Associations
Look in your telephone book under the town or county

Road Safety Officer
England: at your local council offices
N. Ireland: look in the telephone book under 'Department of the Environment'
Scotland: at your regional council offices or ask the local police

Royal Society for the Prevention of Accidents
Cannon House, Priory Queensway, Birmingham B4 6BS
Tel 021 200 2461

St Andrew's Ambulance Association
St Andrew's House, 48 Milton St, Glasgow G4 0HR
Tel 041 332 4031
(Your local branch will be in the telephone book)

St John Ambulance
1 Grosvenor Crescent, London SW1
Tel 071 235 5231
(Your local branch will be in the telephone book)

Scout Association
Baden Powell House, Queens Gate, London SW7 5JS
Tel 071 584 7030

Sports Council
16 Upper Woburn Place, London WC1H 0QP
Tel 071 388 1277

Trading Standards Officer
Look in your telephone book or ask the town hall

Under-fives (or under-eights) adviser
At your local council offices

Index